Secure Detection and Control in Cyber-Physical Systems

In this pioneering reference work, Drs. Yan, Mo, and Wen explain secure detection and control algorithms for cyber-physical systems; describe their history and development, recent advances, and future trends; and provide practical examples to illustrate the topic.

Secure Detection and Control in Cyber-Physical Systems: Working in the Presence of Malicious Data presents readers with the basic concepts of cyber-physical systems, secure detection, and control theory, and explanations of new designs for secure detection and control algorithms that can provide acceptable system performance in the presence of attacks. The authors combine recent research results with a comprehensive comparison of such algorithms and provide ideas for future research. They also give a concise overview of the state-of-the-art cyber-physical system security in a systems and control framework. Content is presented throughout in plain text with equations. Tables and charts are also included to complement the descriptions of the algorithms and aid reader's understanding. Throughout, the authors also present practical examples to illustrate the main ideas. Through this book, readers will gain a comprehensive understanding of the field, including its history, recent advances, and future trends. Readers will be able to apply the relevant algorithms to cyber-physical systems in various contexts – such as aerospace, transportation, power grids, and robotics – and enhance their resiliency to attacks.

This book is vital for researchers and engineers who are researching and working in the fields of cyber-physical systems, secure detection, control theory, and related topics. It is also hugely beneficial for students in the fields of information technology, control systems, or power systems. Readers should have a basic understanding of linear algebra, convex optimization, and stochastic processes.

Secure Detection and Control in Cyber-Physical Systems

Working in the Presence of Malicious Data

Jiaqi Yan, Yilin Mo, and Changyun Wen

CRC Press
Taylor & Francis Group
Boca Raton London New York

CRC Press is an imprint of the
Taylor & Francis Group, an **informa** business

Designed cover image: @Getty, with modifications by authors

First edition published 2026
by CRC Press
2385 NW Executive Center Drive, Suite 320, Boca Raton FL 33431

and by CRC Press
4 Park Square, Milton Park, Abingdon, Oxon, OX14 4RN

CRC Press is an imprint of Taylor & Francis Group, LLC

ISBN: 978-1-032-52922-6 (hbk)
ISBN: 978-1-032-52923-3 (pbk)
ISBN: 978-1-003-40919-9 (ebk)

DOI: 10.1201/9781003409199

Typeset in CMR10
by KnowledgeWorks Global Ltd.

You have to do the best with what God gave you.

— *Forrest Gump*

Contents

Preface

A cyber-physical system (CPS) is a complex system embedding advanced computation, communication, and control techniques into physical spaces, and is usually built up with a set of networked agents such as sensors, actuators, control processing units, and communication devices. Although the development of CPS facilitates efficient and real-time collaboration between elements, the open nature of communication networks makes it rather vulnerable to malicious attacks. Given that the applications of CPSs vary regarding aerospace, transportation, and power grids, which are always safety critical, researchers have acknowledged the importance of designing a system with secure algorithms.

This book first characterizes the properties required for a secure system and possible security threats. Driven by the concerns of deception attacks on communication channels, we are studying secure detection and control in an adversarial environment. New designs on detection and control algorithms will be developed in this book, providing acceptable system performance in the presence of attacks.

Chapter 3 investigates the binary hypothesis testing in an adversarial environment, where a detector determines the true state of an unknown parameter using m sensors. Among these sensors, n out of them can be compromised by the adversary and send arbitrary data. The exponential rate, at which the worst-case probability of detection error goes to 0, is adopted to depict the system performance. This problem is then formulated as a game between detector and attacker, where the former player attempts to maximize this rate and the latter intends to minimize it. We study both cases where $m > 2n$ and $m \leq 2n$, and obtain an equilibrium strategy pair of detection rules and attack schemes for both cases.

Inspired by the fact that the unreliable data transmission can degrade the performance of traditional control algorithms, Chapter 4 discusses the resilient consensus in multi-agent systems, where some of the agents might be misbehaving. Specifically, a continuous-time second-order system is considered, where the agent's dynamics are governed by both position and velocity states. To avoid continuous communication and control, we propose an impulsive secure algorithm. Based on this strategy, signal transmissions and control actions only occur at (aperiodic) sampling instants. After creating a "safe region" with the position states from neighbors, each benign agent derives its control signal with a value inside this region. Sufficient conditions related to the network topology and the maximum number of tolerable faulty nodes are finally derived. As a result, the position states of benign agents are asymptotically synchronized, and the velocity states converge to 0.

Chapter 5 also studies the problem of resilient consensus in multi-agent systems. At this time, we intend to propose secure algorithms that not only facilitate the agreement among benign agents, but also guarantee that the agreement is within the convex hull formed by benign agents' initial states. Toward this end, a resilient consensus algorithm is given, where at each time, the normal agent sorts its received values on one dimension, computes two "middle points" based on the sorted values, and moves its state toward these middle points. An explicit approach is further given for the computation of middle points through linear programming. Compared with the existing works, our approach is applicable to general multi-dimensional systems and introduces lower computational complexity. As the consensus among agents arguably forms the basis of distributed computing, the aforementioned results represent a first step toward the development of secure coordination protocols.

Chapter 6 focuses on another important application of multi-agent systems, namely the resilient containment control in the presence of multiple leaders. Both the leaders and followers can be malicious. In contrast to the leaderless consensus, the objective of this problem is not to achieve an agreement, but to drive the normal followers to the convex hull formed by normal leaders. To this aim, we design secure protocols for both the first-order and second-order systems. Through convex analysis and Lyapunov functions, convergence and resiliency of the proposed algorithms are theoretically proved.

In summary, this book considers the secure detection and control in the presence of deception attacks. All of the proposed approaches are well-supported by numerical examples besides theoretical analysis.

Authors

Jiaqi Yan earned her bachelor's from Xi'an Jiaotong University, China in 2016 and PhD from Nanyang Technological University, Singapore in 2021. She is now a professor in the School of Automation Science and Electrical Engineering, Beihang University. Prior to this, she was a postdoc at Automatic Control Laboratory, ETH Zurich from 2023–2024, a JSPS research fellow with Tokyo Institute of Technology from 2021–2023, a research assistant with Tsinghua University from 2020–2021, and a visiting scholar with California Institute of Technology, USA, in 2019. Her research interests include distributed control, security of cyber-physical systems, and machine learning.

Yilin Mo is an associate professor in the Department of Automation, Tsinghua University. He earned his PhD in electrical and computer engineering from Carnegie Mellon University in 2012 and his BEng from the Department of Automation, Tsinghua University in 2007. Prior to his current position, he was a postdoctoral scholar at Carnegie Mellon University in 2013 and California Institute of Technology from 2013–2015. He held an assistant professor position in the School of Electrical and Electronic Engineering at Nanyang Technological University from 2015–2018. His research interests include secure control systems and networked control systems, with applications in sensor networks and power grids.

Changyun Wen earned a BEng from Xi'an Jiaotong University, China in 1983 and PhD from the University of Newcastle, Australia in 1990. From 1989–1991, he was a postdoctoral fellow at University of Adelaide, Australia. Since 1991, he has been with Nanyang Technological University, Singapore where he is a full professor. His main research activities are in the areas of control systems and applications, cyber–physical systems, smart grids, complex systems, and networks. As recognition of the scientific impact of his publications in these areas, he was listed as a Highly Cited Researcher by Clarivate in 2020, 2021, and 2022.

Prof. Wen is a fellow of IEEE and the Academy of Engineering, Singapore. He was a member of the IEEE Fellow Committee from 2011–2013 and a distinguished lecturer of IEEE Control Systems Society from 2010–2013. Currently, he is the co-editor-in-chief of *IEEE Transactions on Industrial Electronics*, associate editor of *Automatica*, and executive editor-in-chief of *Journal of Control and Decision*. He also served as an associate editor of *IEEE Transactions on Automatic Control* from 2000–2002, *IEEE Transactions on Industrial Electronics* from 2013–2020 and *IEEE Control Systems Magazine* from 2009–2019. He has been actively involved in organizing international conferences playing the roles of general chair (including IECON 2020 and 2023), TPC chair (e.g., TPC chair of Chinese Control and Decision Conference since 2008).

He was the recipient of a number of awards, including the prestigious Engineering Achievement Award from the Institution of Engineers, Singapore in 2005, and the Best Paper Award of IEEE Transactions on Industrial Electronics in 2017.

1

Introduction

This chapter presents introduction of the book. First, we state the background and motivation of this research. Then, major contributions are highlighted, followed by the outline of this book.

1.1 Background

1.1.1 Cyber-Physical System

The cyber-physical systems (CPSs) refer to a new class of engineered systems, serving as integrations of computation, networking, and physical processes [8]. Particularly, the cyber worlds, via embedded computers, control and monitor the physical processes through communication networks (Figure 1.1). Compared with the traditional systems, CPS is typically built up with sets of networked agents: sensors, actuators, control units, etc., and can be regarded as a network of interacting elements with physical inputs and outputs. The ability to interact with physical worlds through computation, communication, and control techniques facilitates efficient and real-time collaboration between elements.

The applications of CPS lie in the sensor-based communication-enabled autonomous systems. Because of growing advancements in embedded systems and communication technologies, more and more systems have revealed the characteristics of CPS. For example, in transportation systems, a wireless sensor network continuously monitors the road condition and transmits the processed information to a computational node to make real-time decisions and control. Other types of CPSs include smart grids, medical monitoring, multi-robot networks, and so on. We list a few of CPS applications in Table 1.1.

1.1.2 Security-Related Issues in CPS

CPS applications involve components that interact through communication networks. Despite the enormous advantages brought by such systems, the open nature of communication networks makes the system much more vulnerable to

DOI: 10.1201/9781003409199-1

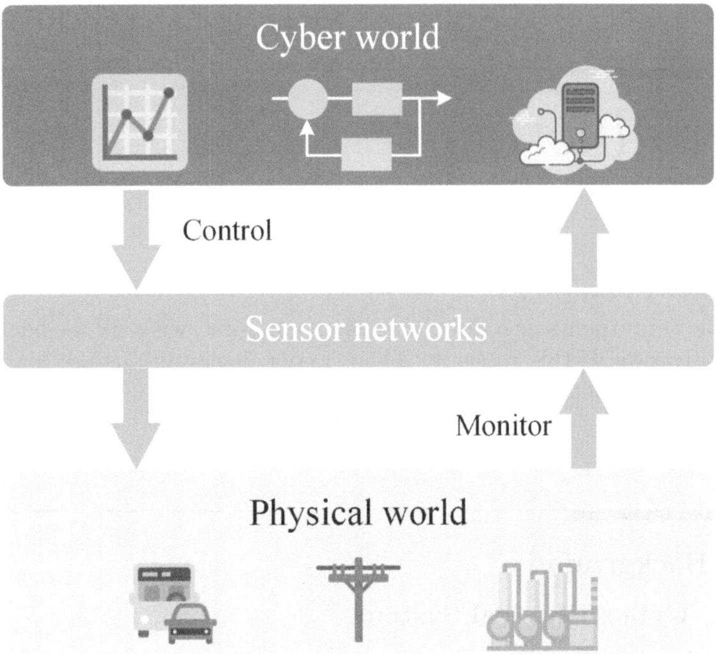

FIGURE 1.1: Cyber-physical systems: The sensor networks collect the monitoring information from the physical processes and transmit it to the computing systems. By dealing with the knowledge coming from the plants, the computing systems, through sensor networks, provide real-time control commands to the physical processes.

TABLE 1.1: Applications of CPSs.

Applications	Issues/Aspects	Reference
Transportation	Design of Cyber-physical vehicle systems	[127], [7], [112]
	Road monitoring	[131], [101]
	Distributed car control system	[85], [70]
Health care	Design of health-care devices	[52], [78],[59]
	Development of medical application platform	[48]
	Early detection on physical abnormality	[46]
Smart buildings	Design of smart bulidings	[75], [144]
	Energy management in smart homes	[143], [65]
Smart grid	Improvement of energy efficiency	[160],[159]
	Energy systems modeling	[57]
	Energy resource management	[109],[43]

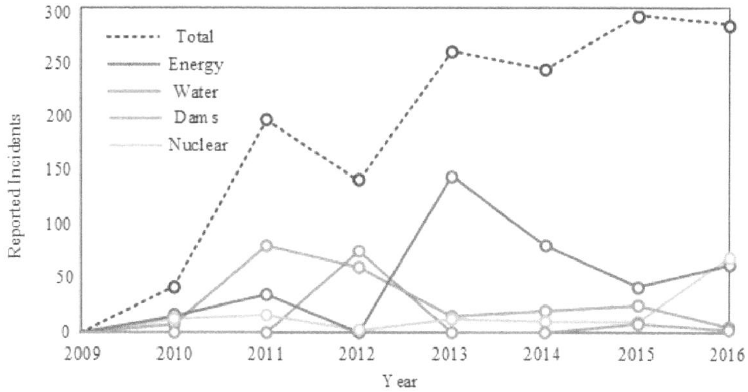

FIGURE 1.2: Reported number of cyber incidents in U.S. received and responded by ICS-CERT with fiscal year [56].

potential cyber attacks. Annual reports from the Industrial Control Systems Cyber Emergency Response Team (ICS-CERT) witness the increasing cyber attacks on control systems (Figure 1.2).

Table 1.2 lists some well-known cyber incidents in history. For many years, malicious attackers have targeted the cyber or communication layers of control systems and managed the critical infrastructures. One concrete example is an incident in 2010 caused by an advanced computer worm, Stuxnet. By targeting the Supervisory Control and Data Acquisition (SCADA) systems, Stuxnet destroyed the energy facilities, and was particularly responsible for causing substantial damage to Iran's nuclear program. Another instance is the Ukraine Power System Attacks in 2015–2016. At least three power regions in Ukraine were announced to be compromised by cyber attackers. By invading the monitoring and management systems, this attack resulted in failures of seven substations of 110 KV and 23 substations of 35 KV, and led to hours of blackouts in more than half of the region. More details of these two attacks are demonstrated in Figures 1.3 and 1.4, adopted from [68] and [1], respectively.

The above cyber incidents highlight potential threats to control systems. Since most of the CPS applications are safety-critical, the failure of these systems could lead to large economic losses and even cause irreparable harm to public health [19]. Thus more and more research attention has been paid to the security of CPS.

1.1.3 Security Goals and Threats

Information security focuses on three objectives: confidentiality, integrity, and availability, which are collectively known as CIA, as shown in Figure 1.5. Specifically, confidentiality refers to the ability to keep the information secret from unauthorized ones. Integrity is to ensure the accuracy and

TABLE 1.2: Some cyber incidents in history.

Year	Name	Description
2010	Stuxnet	The world's first publically known digital weapon, causing substantial damage to Iran's nuclear program.
2013	New York Dam	An Iranian hactivist group launched a cyber attack against the control system for Bowman Dam in New York.
2014	Citibank	By launching cyber attacks on Citibank, the attacker has stolen tens of millions of dollars.
2014	German Steel Mill	A steel mill in Germany suffered a cyber attack, which resulted in massive damage to the system.
2015	Ukraine Power Grid Attack No. 1	The first known successful cyber-attack on a country's power grid.
2016	"Kemuri" water company	By intruding into the programmable logic circuits (PLCs), the attackers manipulated control applications and altered water treatment chemicals.
2016	Ukraine Power Grid Attack No. 2	Cyber-attackers tripped breakers in 30 substations, causing blackout incidents which affected over 225,000 customers in the region of Lavno-Franklvst of Ukraine.
2017	Uber Data Breach	Hackers accessed a server containing personal information of more than 57 million Uber drivers and riders. They demanded a $100,000 ransom to delete their copy of data.
2019	Venezuela Power Outage	The populace in Venezuela suffered a power blackout which is suspected to have been caused by hackers backed by U.S. Intelligence.

trustworthiness of data. Availability, on the other hand, guarantees that the data, network resources, or services are continuously available to the legitimate users whenever they need them.

In general, an adversary, by taking control of sensors or actuators in the communication networks, can deteriorate the system security in terms of confidentiality, integrity, or availability. Accordingly, Cardenas et al. classify the cyber attacks against CPSs into three types: eavesdropping, deception, and denial-of-service (DoS) attack [20], which are detailed below:

- Eavesdropping attacks: An adversary infers the states of systems by taking advantage of unsecured communications to access the transmitted data. The systems under this attack will suffer privacy disclosure. For example, a healthcare CPS requires the patients' health information to be transmitted to the doctors. Yet, by using the internet-connected medical devices, users

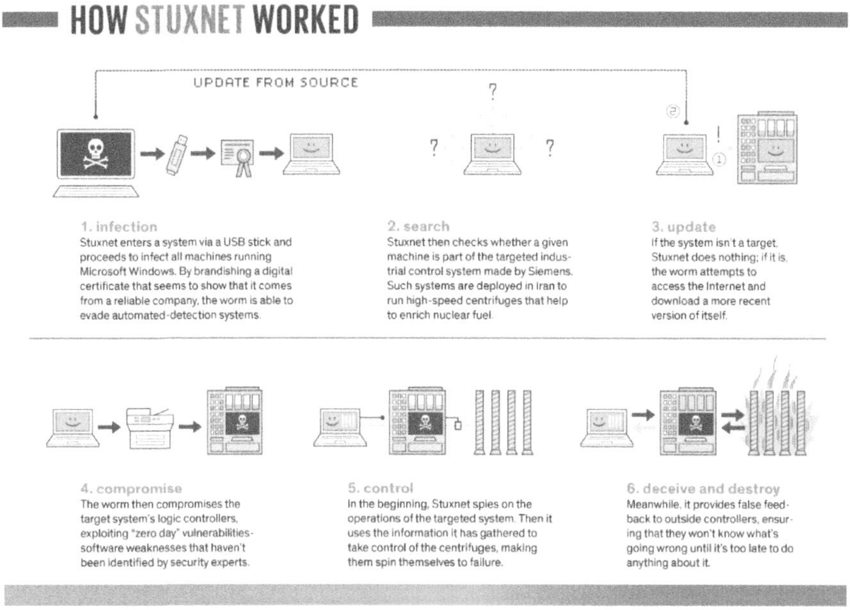

FIGURE 1.3: The attack procedure of Stuxnet[68].[1]

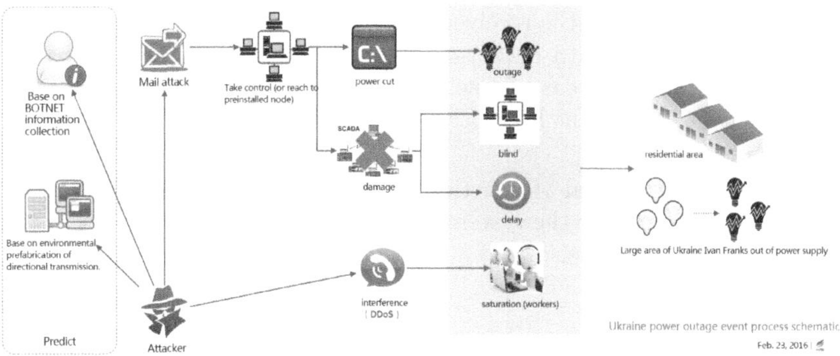

FIGURE 1.4: The summary of Ukraine power outage attacks [1].

may unwittingly expose their sensitive data to an attacker. This attack is difficult to be detected as the systems appear to be operating normally.

- Deception attacks: Also known as integrity attacks, where an adversary maliciously modifies the transmitted data. The receiver, after receiving this false data, is fooled into believing this incorrect version of reality to

[1]Reproduced with permission of ©2013 IEEE.

FIGURE 1.5: CIA Triangle: Three goals in information security.

be true and act in a way that benefits the attacker. Because the conflict of interests among parties is almost inevitable, deception attacks are commonly encountered in systems. Some of the common deception attacks include replay attack [99] and false data injection attack [83].

- DoS attacks: The attacker intentionally jams or blocks the communication channels so that the receiver fails to receive certain packets from the sender. Compared with deception attacks, DoS attacks are more realizable because of their easy implementation and limited system knowledge requirement. Abundant historical cyber incidents highlighted the hazard of DoS attacks.

Figure 1.6 offers a general abstraction of security threats in CPS, where the adversary can deteriorate the system performance by launching cyber attacks on communication channels.

1.2 Contributions of the Book

Inspired by the above security concerns, particularly the ones induced by deception attacks, this book focuses on secure detection and control in cyber-physical systems, with the aim of designing resilient algorithms or architectures that survive CPS attacks. The main contents and contributions of this book are highlighted as follows:

- *Secure Detection in Wireless Sensor Networks (Chapter 3)*:
  Wireless sensor networks (WSNs) are key components in the emergent CPSs. They are particularly deployed as interfaces through which data

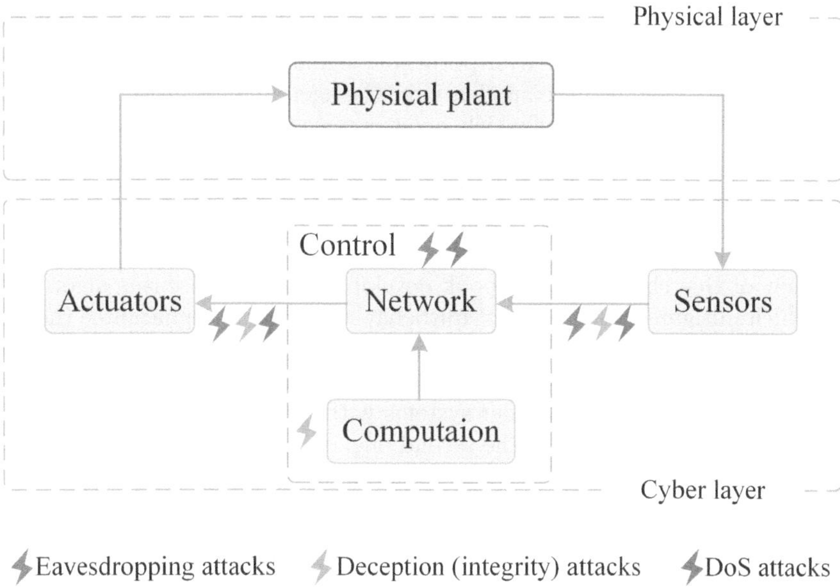

FIGURE 1.6: A general abstraction of cyber security threats in CPS [35].[2]

are collected from the physical layer and transferred to the cyber layer, as well as interfaces through which commands/instructions are injected from the cyber layer to the physical layer. WSNs may include hundreds of spatial sensors which interact to solve complex tasks such as detection, estimation, and control. However, at the same time, the extensive use of sensors also makes networks vulnerable to potential attacks, such as message manipulations, false data injections, etc. Driven by such security concerns, this book considers secure detection in WSNs, where a detector determines the true state of an unknown parameter through the measurements of multiple sensors. Given that an attacker maliciously modifies some sensor's measurements, the detection performance can be easily compromised. To address this issue, a secure detector is presented which achieves the optimal detection performance under attacks. By formulating this problem as a game between the detector and attacker, an optimal attack strategy is also developed, which forms a Nash equilibrium with the proposed optimal detector.

- *Resilient Consensus in Adversarial Environment (Chapters 4 and 5)*: CPSs require the interaction among different components. Multi-Agent System (MAS), which has emerged over the years, is one of the major technological paradigms regulating interactions among autonomous

[2]Reproduced with permission from ©2019 Elsevier.

entities. As such, it has the potential to support CPS in implementing a highly distributed architecture [14]. The concept of MAS has been widely used in applications like transportation, signal processing, and sensor networks. In many of these applications, the agents must agree upon certain quantities of interest, at which point a consensus is said to be achieved. However, most existing algorithms to facilitate consensus are rather fragile: their performance will be greatly degraded due to the proliferation of "misbehaving" agents, i.e., the agents whose information is manipulated by malicious attacks or who refuse to follow the prescribed algorithms. In view of this challenge, this book develops resilient consensus algorithms, which mitigate the impacts of misbehaving agents and guarantee the consensus of the rest benign ones. Particularly, we present two algorithms. The first one, through impulsive control, facilitates the resilient consensus in continuous-time second-order systems with aperiodic communication signals and control actions. The second one, on the other hand, achieves the resilient agreement within the convex hull of benign agents' initial states. In comparison with the existing works, one major breakthrough of this approach is that the applicability has been extended to multi-dimensional spaces with reduced computational cost. Since the consensus arguably forms the foundation for distributed computing, these results lay a solid foundation for future works to develop resilient coordination protocols.

- *Resilient Containment Control in Adversarial Environment (Chapter 6)*: The existing consensus algorithms often focus on the leaderless scenarios. Yet, in practice, there might be the case that one or more leaders exist among these agents. The single leader case is well studied in the leader-following consensus problems. Containment control, on the other hand, comes from the existence of multiple leaders in MASs, where the followers aim to move towards the convex hull spanned by multi-leaders. This book further investigates the resilient containment control in adversarial environments, where both the leaders and followers can be misbehaving. We propose secure protocols for both the first-order and second-order systems. Through convex analysis and Lyapunov functions, convergence and resiliency of the proposed algorithms are theoretically proved. Namely, they guarantee that the benign followers eventually move into the convex hull formed by benign leaders, in spite of the network misbehaviors. To the best of our knowledge, this is the first work regarding secure containment control.

1.3 Outline of the Book

The remainder of this book is organized as follows:

Chapter 2 reviews the related works to CPS security, including the existing approaches in information security and automatic control. Some technical preliminaries are also provided in this chapter, which would be useful in the analysis required in subsequent chapters.

In Chapter 3, the sequential detection in WSNs is investigated, where the detector determines the true state of an unknown parameter based on the measurements from m sensors, and the attacker deteriorates the detection performance by compromising $n\,(\leq m)$ sensors' measurements. The exponent rate, at which the worst-case probability of detection error goes to 0, is adopted as the performance metric. This problem is formulated as a game between the detector and attacker. We study both cases where $m > 2n$ and $m \leq 2n$. In each case, certain optimal strategies are proposed for both players, with which a Nash equilibrium is achieved.

Chapters 4–6 consider the resilient distributed control in MASs. Particularly, the leaderless consensus problem is discussed in Chapters 4 and 5, aiming at achieving consensus under malicious nodes. Towards this end, Chapter 4 proposes an impulsive resilient consensus algorithm in continuous-time second-order systems, allowing the synchronization of position states through aperiodic communication and control signals.

In Chapter 5, a "middle points"-based algorithm is alternatively proposed, which is also applicable to multi-dimensional systems. By solving middle points through a linear programming, the proposed strategy introduces a reduced computational complexity in high-dimensional spaces. Despite the presence of misbehaviors, it not only ensures that the benign agents exponentially reach an agreement but also improves the consensus accuracy by guaranteeing that the agreement is within the convex hull formed by benign agents' initial states.

Chapter 6 discusses another important application in MASs, i.e., containment control in the presence of multiple leaders. In practice, misbehaving agents may exist, which deteriorate the system performance by either misleading the normal followers to leave the convex hull formed by leaders or destroying the convex formation of leaders. To deal with this issue, resilient algorithms are respectively presented for first-order and second-order systems, driving the normal followers into the convex hull spanned by normal leaders. Under certain topological conditions, both algorithms are proved to be resilient enough to tolerate a number of malicious nodes.

Finally, Chapter 7 summarizes the book and presents several new perspectives regarding CPS security.

2

Literature Review

The security of CPS relies on the integration of cyber and physical layers, as well as the different ways they are affected by decision-makers [122]. Therefore, this field lies in the intersection of computer science and control systems. In this chapter, an in-depth survey of the existing techniques on related topics, including information technology (IT) security and secure control, is presented.

2.1 Related Work in Information Security

CPS applications arise in critical infrastructures like transportation, environment, and electricity grids. Such systems are typically built on IP/TCP-based communication networks. Therefore, the first question would be: can CPS security be handled purely with IT solutions?

An important tool in information technology for securing systems is authentication. It enables organizations to protect their resources, including computer systems, databases, websites, and other network-based applications or services, by ensuring that only authenticated users (or processes) can get access to them.

Historically, biometrics have been used as the most common method for authentication [12]. Usual biometric authentication methods include fingerprint identification, voice analysis, face recognition, et al. There are a number of advantages to biometric authentication. It can provide accurate, secured access to information, as fingerprints, voice, and iris represent absolutely unique data sets. Meanwhile, these identities can be verified without resorting to documents that might be stolen or lost.

Secure communications between authorized parties can also be guaranteed with the help of cryptographic methods. One mature solution is digital signature; see Figure 2.1. To achieve it, each entity should store the public keys for others and the private key for its own. Apart from the plaintext, the sender also transmits the digital signature to the receiver by the following operations: The plaintext is processed using hash functions to generate a message digest; the sender then uses its private key to encrypt the message digest, forming a digital signature that can be decrypted with its public key. Upon

DOI: 10.1201/9781003409199-2

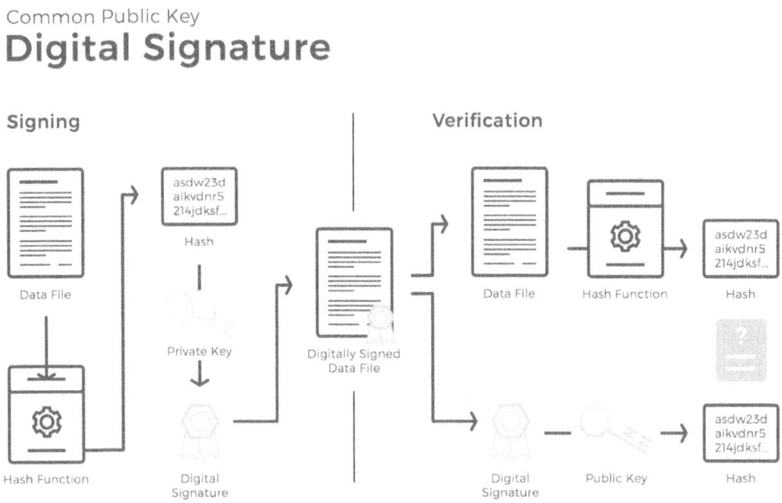

FIGURE 2.1: An overview of the concept of digital signature [2].

receiving the message, data decryption and verification processes are hence required: the receiver first hashes the received plaintext into message digest 1, decrypts the message digest 2 from the digital signature using the sender's public key, and finally compares the two digests to verify the information ([96]). A digital signature gives a recipient strong confidence to believe that the message was sent from a legal sender and has not been tampered by a third party.

Other security tools developed by IT society include challenge/response mechanisms (where one side presents a challenge to be answered and the other side presents a correct answer to the challenge to get authenticated [28]) and trusted timestamps (which add extra security to digital signatures by keeping secure track of the creation and modification time of a document [31]). The aforementioned authentication and verification tools can test the integrity of messages and thus limit the vulnerability to cyber attackers.

Incorporating the traditional IT methods into system design, such as encryption and authentication of transmitted messages, is important. However, it cannot serve as a full solution to CPS security [122]. On one hand, many encryption algorithms are computationally expensive, especially for embedded systems [119]. As a consequence, they are likely to introduce large time delays to the closed-loop system. Furthermore, the above mechanisms can often be subverted by inevitable human errors and design flaws, which create vulnerabilities for external adversaries [19]. For example, a birthday attack can abuse the communication between parties by taking advantage of the collision of hash functions [11, 45]. Therefore, even if certain communication channels have been encrypted, malicious attackers can never be completely ruled out and may lead to undesired actions of the controller. This is especially undesirable as most of the CPS applications are safety-critical: they must provide acceptable system performance even under attacks. On the other hand, IT security mainly focuses on the protection of information at the cyber layer of systems. As shown in Figure 1.1, the communication networks of CPSs enable the deep coupling of cyber and physical worlds, imposing fundamental challenges for the pure cyber security tools. Particularly, feedback loops inherent to CPSs imply the interdependence of the security of cyber and physical worlds. It is thus desired to investigate how attacks at the cyber layer affect the estimation and control performance at the physical layer. In particular, control engineers aim at improving the system security by taking advantage of the dynamics of physical plants.

In summary, the topic of cyber attacks is not of interest just to the IT security community but must be studied from a comprehensive system and infrastructure perspective. Researchers therefore argue for the need to draw new design and analysis tools from control theory.

2.2 CPS Security from Control Perspective

The feedback loops in CPSs have implications on the underlying physical dynamics and highlight the study of cyber security from a system and control point-of-view. Particularly, CPS security should pay its attention not only to the underlying communication networks but also to the sensors and actuators forming feedback loops. In this section, we shall present a brief review of the control-oriented secure measures, as demonstrated in Figure 2.2.

2.2.1 Prevention

Prevention mechanisms have been proposed in literature to avoid the information leakage of dynamic systems. Dibaji et al. [35] classify the existing methods into two classes: *cryptography* and *randomization*.

Cryptography, as detailed in Section 2.1, is a long-standing research of interest in computer science and information security. Randomization, on the other hand, is a defensive tool that aims at confusing the potential attackers by introducing randomness to the true states. It is especially useful when the adversaries can leverage the deterministic updating rules to predict and obtain key information about the systems.

Differential privacy (DP) [41] is one of the standard approaches using randomization. In the last decades, researchers have resorted to DP in the design of privacy-preserving algorithms with the aim of protecting various objectives,

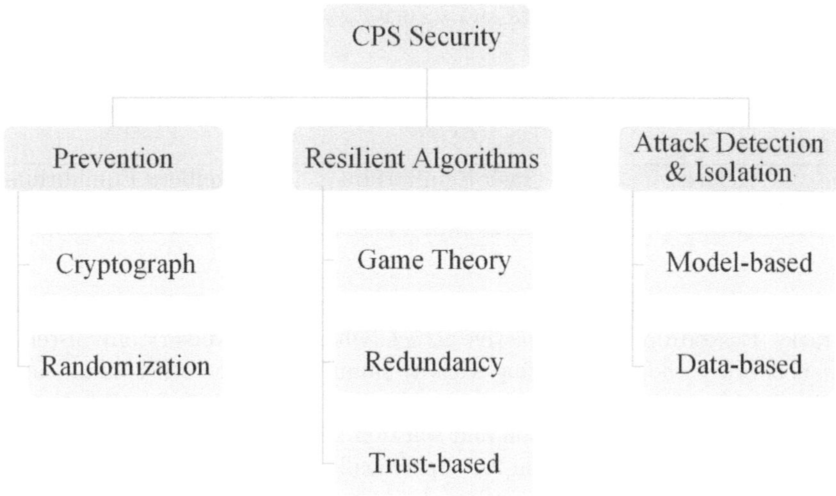

FIGURE 2.2: Classification of security measures from system and control perspective.

including control commands [55, 141], initial states [54, 105], network topology [121], and training data in distributed machine learning [139, 25]. Other schemes based on randomization can be found in [98, 49]. For example, Mo et al. [98] propose an average consensus algorithm to guarantee the privacy of initial states and asymptotic consensus on the exact average of the initial values by masking the transmitted data with well-designed random noises. The idea of randomization has also been proposed in the setting of adversarial machine learning [53]. Generally, there exists a trade-off between systems' utility and privacy. Works including [123, 132, 88] explore this trade-off as well.

2.2.2 Resilient Algorithms

System resilience is an ability of the system to withstand a disruption within acceptable degradation parameters and continue to carry out its mission in the face of adversity [118, 150]. Resilience may not be an inherent attribute of the system but can be established through the dedicated design of control and optimization strategies.

Resilience-increasing algorithms have been widely studied in the control community. As an important measure, the game-theoretic approach has been successfully applied to model the interplay between the system operator and attacker. The defender provides system resilience by maximizing the cost for the attacker to deteriorate the system [26, 42, 152] or minimizing the damage that an attacker can introduce to the system [116, 36, 76]. Depending on the model of players, different game strategies have been proposed. For example,

TABLE 2.1: Solution concepts and security game scenarios between attacker and defender [89].

Attacker \ Defender	Active	Passive
Active	Nash Equilibrium	Stackelberg Equilibrium
Passive	Stackelberg Equilibrium	Nash Equilibrium

communication channels may suffer from eavesdropping and jamming (DoS) attacks. Eavesdropping is a passive attack, where the adversary only listens to the network without interacting with it. Jamming, on the other hand, can be active by blocking the channels such that the operator is impossible to communicate with. The interaction and solution concepts between passive/active players are captured in Table 2.1. Specifically, if an eavesdropper only passively receives the "leaking" information from channels, it can be modeled as a follower in a Stackelberg game [44] playing against a defender who is a leader and employs active measures. Similarly, the interplay between an active attacker and a passive defender is also reasonably viewed as a Stackelberg game. On the other hand, in the scenario where both players behave actively or either side has an information advantage, the Nash equilibrium [91] becomes a reasonable solution concept. In addition, other approaches from a game-theoretic view have been proposed as well. For instance, non-cooperative games are formulated and addressed in multi-agent systems [156]. In [116], a zero-sum game is suggested for detection under deception attacks.

Apart from the game-theoretic methods, a group of works increases system resilience through redundancy: deploying additional nodes and channels to enhance the network robustness [73, 130, 33, 154, 153]. These methods are proved efficient, especially in multi-agent systems, as the redundant agents and links "compensate" for the attacking or faulty signals on misbehaving ones and will be further discussed in Chapter 5. Moreover, system resilience can alternatively be enhanced with trust-based approaches by protecting the availability and integrity of a small subset of elements and guaranteeing the correct information spread along the paths formed by these trusted elements [3, 117, 4, 161, 155]. Since device hardening is costly, the set of trusted components must be small and deployed at crucial points. Particularly, Abbas et al. [4] show if the set of trusted agents induce a connected dominating set, the network is able to tolerate any number of misbehaving nodes and achieve resilient consensus.

2.2.3 Attack Detection and Isolation

Finally, attack detection and isolation has received considerable attention from researchers, which usually identifies and removes the existence of an attack by monitoring its influence on the outputs of CPSs.

One commonly adopted strategy from control community is to construct an observer-based detection mechanism, by taking into account the dynamics of physical systems and addressing how this model can be used to identify the anomaly [111, 93]. With this mechanism, one estimates the system states with an observer, and make decision on possible misbehaviors based on the residuals generated by the estimation. One concrete example is Bad Data Detection (BDD) of power systems [30, 22, 102]. The system seeks estimates of the state variables that best fit the meter measurements. As errors could enter the meter measurements due to device failures, malicious attacks, etc., BDD is introduced to protect the state estimation. It calculates the measurement residual and compare its Euclidean norm with a prescribed threshold. Once the norm exceeds this threshold, an alarm will be triggered. Given that an attacker can inject arbitrary signals into the system, some works consider malicious data as an unknown input with no *a priori* knowledge. An unknown input observer (UIO) is thus designed, ensuring the estimation always track the actual state, regardless of the value and distribution of the unknown inputs [23, 111, 93, 148, 149]. This fact enables UIO to figure out the attacking signals based on the system models. Besides, Luenberger observers are also widely used [40, 87].

The concept of physical watermarking also emerges in the context of detection and isolation, especially for replay attacks [99, 63, 62]. In replay attacks, the adversary records a sequence of system outputs and later replay them to the operator. If the system is operating in steady state, these replayed signals will be statistically identical to the outputs of the system under normal operation, posing significant challenges to the classical detection rules. To address this issue, Mo et al. [100] develop a detection method by injecting the well-designed noisy input, termed as watermarked signal, to physical systems. As the attacker is unaware of this physical watermark, the system cannot be adequately emulated and hence achieves improved detection performance.

The aforementioned model-based approaches is enabled by its ability to recognize the abnormities in system dynamics. On the other hand, data-based methods have also been popular. By posing the intrusion detection as statistical learning problems, machine learning algorithms, such as support vector machines, self-organizing maps, Bayesian networks, etc, are used in literature to classify the measurements as being either secure or attacked [108, 120, 61].

2.3 Technical Preliminaries

After reviewing the related work on CPS security, this subsection introduces some useful technical preliminaries, which would be applied in the rest of book.

2.3.1 Graph Theory

This section reviews some fundamentals of graph theory. When dealing with MASs, it is common to model the communication network with a graph $\mathcal{G} = \{\mathcal{V}, \mathcal{E}\}$, where $\mathcal{V}$ is the set of agents, and $\mathcal{E} \subset \mathcal{V} \times \mathcal{V}$ is the set of edges, modeling the information flow or influence between agents, and typically realized through communication or sensing. The edge $e_{ij} \in \mathcal{E}$ indicates that agent i can directly receive information from agent j. Simple graphs satisfy that $e_{ii} \notin \mathcal{E}, \forall i \in \mathcal{V}$.

A graph is said to be undirected if and only if $e_{ij} \in \mathcal{E}$ implies $e_{ji} \in \mathcal{E}$. The neighborhood of an agent $i \in \mathcal{V}$ is then defined as $\mathcal{N}_i = \{j \in \mathcal{V} | e_{ij} \in \mathcal{E}\}$. Otherwise, the graph is referred to be a directed graph (or digraph). In this case, the set of in-neighbors and out-neighbors of agent i might be different, and are respectively defined as $\mathcal{N}_i^+ \triangleq \{j \in \mathcal{V} | e_{ij} \in \mathcal{E}\}, \mathcal{N}_i^- \triangleq \{j \in \mathcal{V} | e_{ji} \in \mathcal{E}\}$.

2.3.2 Network Robustness

Network robustness is a connectivity measure for graphs. Specifically, it formalizes the notion of sufficient local redundancy of information flow in the network, and thus is useful for the study of resilient distributed algorithms which use only local information [71].

The notions of network robustness are based on the r-reachability as defined below:

Definition 2.1 **(r-reachable)** *In a graph $\mathcal{G}(\mathcal{V}, \mathcal{E})$, a subset $\mathcal{S} \subseteq \mathcal{V}$ is said to be r-reachable if it contains a vertex that has at least r (in-)neighbors from outside $\mathcal{S}$. That is, there exists $i \in \mathcal{S}$ such that $\mathcal{N}_i \backslash \mathcal{S}$* [1].

Based on this notion, network robustness, first introduced in [72], is formally defined as follows:

Definition 2.2 **(r-robust)** *A network modeled by $\mathcal{G} = \{\mathcal{V}, \mathcal{E}\}$ is said to be r-robust, if for any pair of disjoint and nonempty subsets $\mathcal{V}_1, \mathcal{V}_2 \subsetneq \mathcal{V}$, at least one of the sets is r-reachable.*

The notion of r-robustness can be further generalized as follows: for $r \in \mathbb{Z}^+$, let $\mathcal{X}_\mathcal{S}^r \subseteq \mathcal{S}$, such that each agent in $\mathcal{X}_\mathcal{S}^r$ has at least r neighbors outside of $\mathcal{S}$, namely

$$\mathcal{X}_\mathcal{S}^r = \{i \in \mathcal{S} : |\mathcal{N}_i \backslash \mathcal{S}| \geq r\}. \tag{2.1}$$

Definition 2.3 **((r, s)-robust)** *A network modeled by $\mathcal{G} = \{\mathcal{V}, \mathcal{E}\}$ is said to be (r, s)-robust, if for any pair of disjoint and nonempty subsets $\mathcal{V}_1, \mathcal{V}_2 \subsetneq \mathcal{V}$, at least one of the following statements holds:*
1) $|\mathcal{X}_{\mathcal{V}_1}^r| = |\mathcal{V}_1|$;
2) $|\mathcal{X}_{\mathcal{V}_2}^r| = |\mathcal{V}_2|$;
3) $|\mathcal{X}_{\mathcal{V}_1}^r| + |\mathcal{X}_{\mathcal{V}_2}^r| \geq s$.

[1] In directed graphs, we replace $\mathcal{N}_i$ with $\mathcal{N}_i^+$.

Clearly, a $(r+s-1)$-robust graph is (r,s)-robust as well. The following lemma also shows the basic properties of the robust graphs:

Lemma 2.1 ([72]) *For a (r,s)-robust graph $\mathcal{G} = \{\mathcal{V}, \mathcal{E}\}$ with $|\mathcal{V}| = N$, the following holds:*
(a) $\mathcal{G}$ is (r', s')-robust, where $0 \leq r' \leq r$ and $1 \leq s' \leq s$, and in particular, it is r-robust.
(b) $r \leq \lceil N/2 \rceil$, where $\lceil \cdot \rceil$ is the ceiling function. Also, if $\mathcal{G}$ is a complete graph, then it is (r', s)-robust for all $0 < r' \leq \lceil N/2 \rceil$ and $1 \leq s \leq N$

In [4], the authors extend the notions of robustness by incorporating trusted nodes. Specifically, given a graph $\mathcal{G} = \{\mathcal{V}, \mathcal{E}\}$, let $\mathcal{T} \subseteq \mathcal{V}$ be a set of trusted nodes. Then, given a non-empty subset $\mathcal{S} \subseteq \mathcal{V}$, define $\mathcal{Y}_\mathcal{S} \subseteq \mathcal{S}$, such that each agent in $\mathcal{Y}_\mathcal{S}$ has at least one trusted neighbor outside of $\mathcal{S}$, that is,

$$\mathcal{Y}_\mathcal{S} = \{i \in \mathcal{S} : (\mathcal{N}_i \backslash \mathcal{S}) \cap \mathcal{T} \neq \varnothing\}.$$

Then let us denote $\mathcal{Z}_\mathcal{S}^r$ as

$$\mathcal{Z}_\mathcal{S}^r = \mathcal{X}_\mathcal{S}^r \cup \mathcal{Y}_\mathcal{S},$$

where $\mathcal{X}_\mathcal{S}^r$ is defined in (2.1). Note that $\mathcal{Z}_\mathcal{S}^r$ is the subset of nodes in $\mathcal{S}$ with each of the agents in $\mathcal{Z}_\mathcal{S}^r$ has either at least r neighbors outside of $\mathcal{S}$, or at least one trusted neighbor outside of $\mathcal{S}$. Moreover, we say that a set $\mathcal{S}$ is r-reachable with $\mathcal{T}$ if the corresponding $\mathcal{Z}_\mathcal{S}^r$ is non-empty. Now, let us define the notions of r-robustness and (r,s)-robustness with $\mathcal{T}$ as follows [4]:

Definition 2.4 (r-robust with $\mathcal{T}$) *A network modeled by $\mathcal{G} = \{\mathcal{V}, \mathcal{E}\}$ is said to be r-robust with $\mathcal{T}$, if for any pair of disjoint and nonempty subsets $\mathcal{V}_1, \mathcal{V}_2 \subsetneq \mathcal{V}$, at least one of the sets is r-reachable with $\mathcal{T}$.*

Definition 2.5 ((r,s)-robust with $\mathcal{T}$) *A network modeled by $\mathcal{G} = \{\mathcal{V}, \mathcal{E}\}$ is said to be (r,s)-robust with $\mathcal{S}$, if for any pair of disjoint and nonempty subsets $\mathcal{V}_1, \mathcal{V}_2 \subsetneq \mathcal{V}$, at least one of the following statements hold:*
1) $|\mathcal{Z}_{\mathcal{S}_1}^r| = |\mathcal{S}_1|$;
2) $|\mathcal{Z}_{\mathcal{S}_2}^r| = |\mathcal{S}_2|$;
3) $|\mathcal{Z}_{\mathcal{S}_1}^r| + |\mathcal{Z}_{\mathcal{S}_2}^r| \geq s$;
4) $(\mathcal{Z}_{\mathcal{S}_1}^r \cup \mathcal{Z}_{\mathcal{S}_2}^r) \cap \mathcal{T} \neq \varnothing$.

Moreover, in many applications of MASs, it is desired to drive the states of some agents (followers) towards the states of another group of agents (leaders). To characterize the network's capability of tolerating the misbehaving nodes in such scenarios, Usevitch et al. [134] give the notion of network robustness with respect to a certain subset $\mathcal{S}$, where $\mathcal{S} \subseteq \mathcal{V}$ is a nonempty subset of $\mathcal{V}$:

Definition 2.6 (strongly r-robust w.r.t. $\mathcal{S}$) *A network modeled by $\mathcal{G} = \{\mathcal{V}, \mathcal{E}\}$ is said to be strongly r-robust w.r.t $\mathcal{S}$ if for any nonempty subset $\mathcal{S}' \subseteq \mathcal{V} \backslash \mathcal{S}$, $\mathcal{S}'$ is r-reachable.*

Intuitively, the network robustness is a connectivity measure for graphs. It claims that for any two disjoint and nonempty subsets of agents, there are "many" agents within these sets that have a sufficient number of neighbors outsides. Compared with the classical k-connectivity [2], the notion of network robustness is more pertinent to quantify the local connectivity of nodes, and thus is adopted widely in characterizing the network topology of achieving local-information-based algorithms. We would also use these notions to evaluate the performance of distributed algorithms proposed in Chapters 4–6.

2.3.3 Sarymsakov Matrix

Another important tool would be the Sarymsakov matrix. Given a row stochastic matrix $A = (a_j^i)$, define the directed graph $\mathcal{G}(A)$ associated with it as $\mathcal{G}(A) = \{\mathcal{V}, \mathcal{E}\}$, where $(i, j) \in \mathcal{E}$ if and only if $a_j^i > 0$. Given any row stochastic matrix $A = \{a_{ij}\}$, its associated digraph is defined as $\mathcal{G}(A) = \{\mathcal{V}, \mathcal{E}\}$, where $e_{ij} \in \mathcal{E}$ if and only if $a_{ij} > 0$. We first have the below result:

Lemma 2.2 ([114]) *The digraph $\mathcal{G}(A)$ has a spanning tree if and only if A has a simple eigenvalue $\lambda = 1$.*

For a set $\mathcal{V}' \subset \mathcal{V}$, its *one-stage consequent indice* [125] is defined by

$$F_A(\mathcal{V}') = \{j : a_j^i > 0 \text{ for some } i \in \mathcal{V}'\}.$$

Namely, $F_A(\mathcal{V}')$ is the set of nodes who have influence on the ones in $\mathcal{V}'$.

Based on the one-stage consequent indices, the Sarymsakov matrix is formally defined below [145]:

Definition 2.7 *(Sarymsakov matrix): A row stochastic matrix A is said to be Sarymsakov, if for any disjoint nonempty sets $\mathcal{V}_1, \mathcal{V}_2 \subsetneq \mathcal{V}$, one of following statements hold:*
1) $F_A(\mathcal{V}_1) \cap F_A(\mathcal{V}_2) \neq \varnothing$;
2) $F_A(\mathcal{V}_1) \cap F_A(\mathcal{V}_2) = \varnothing$ and $|F_A(\mathcal{V}_1) \cup F_A(\mathcal{V}_2)| > |\mathcal{V}_1 \cup \mathcal{V}_2|$.

Namely, Sarymsakov matrix means that, either $\mathcal{V}_1$ and $\mathcal{V}_2$ have some common influencing nodes, or the total number of their influencing nodes is greater than that of being influenced. The below results provide some important properties of Sarymsakov matrices:

Lemma 2.3 ([145]) *Let $\mathcal{A}$ be a set of stochastic matrices. For each sequence of matrices $A_1, A_2, \ldots$ from $\mathcal{A}$, if there is an integer $\alpha \geq 1$ such that for each $k \geq \alpha$ and any $A_i \in \mathcal{A}$, the matrix $A_k \ldots A_2 A_1$ belongs to the Sarymsakov class, then $A_k \ldots A_2 A_1$ converges to a rank-one matrix $\mathbf{1}c^T$ as $k \to \infty$.*

Lemma 2.4 ([15]) *For any Sarymsakov matrix A, its associated graph $\mathcal{G}(A)$ is rooted. On the other hand, if $\mathcal{G}(A)$ is rooted and has a self-arc at each vertex, A is Sarymsakov.*

---

[2] A network is with k-connectivity, if it remains connected after removing any $k-1$ nodes from the network.

3

Sequential Detection with Byzantine Sensors

3.1 Introduction

Wireless sensor network (WSN) is a vital part of CPS, as strong sensing capability is one of the major driving factors for CPS applications. This chapter considers the binary hypothesis testing in WSN, where a detector determines the true state of an unknown parameter through the measurements of m sensors. On the other hand, an attacker aims at degrading the detection performance by modifying some sensors' measurements. In practice, it may get access to the sensors and send arbitrary messages or break the communication channels between the sensors and detector to tamper with the transmitted data. Such sensors, whose information is fully controlled by the adversary, are called Byzantine sensors. Due to the limited resources of the adversary, we assume that at most n out of these m sensors are Byzantine.

According to Kerckhoffs's principle [126], i.e., the security of a system should not rely on its obscurity, we assume that the adversary knows exactly the hypothesis testing algorithm used by the detector. On the other hand, the detector only knows the number of Byzantine sensors n, but does not know the exact set of the compromised sensors. The exponential rate, at which the probability of detection error goes to zero, is adopted to indicate the performance of the detector. We formulate this problem as a game between the detector and attacker, in which the former attempts to maximize this rate, while the latter intends to minimize it. We respectively investigate the cases where $m > 2n$ and $m \leq 2n$, and propose optimal strategy pairs for both players to achieve a Nash-equilibrium. The efficiency of proposed detector in the absence of attacker is finally discussed.

DOI: 10.1201/9781003409199-3

3.2 Related Work

Detection under Byzantine attack has been extensively studied in literatures (see [21, 90, 97, 60, 128]). For example, in [9], Bayram et al. propose a restricted Neyman-Pearson (NP) framework for composite hypothesis testing in the presence of prior distribution uncertainty. The optimal decision rule according to the restricted NP criterion is analyzed in their work. In [90], the authors took the respective of an intruder and found the optimal attacks to minimize the Kullback-Leibler (K-L) divergence of the manipulated measurements.

Moreover, different from [60, 128, 90], where the manipulated measurements are assumed to be independent, this chapter assumes that the Byzantine sensors may collude with each other. We believe this collusion model is more general and realistic, although adding analysis complexity. What's more, instead of making extra assumption on the false information from malicious sensors, like [5, 138], we suppose the malicious data can take any value. Similar attack model can also be found in [97]. However, the authors only focus on one-step detector, while we consider an infinite time sequence of detectors, which is more challenging. Besides, the problem in this chapter is formulated in a game-theoretic way, while the aforementioned works take the perspective of either an attacker or a system manager.

Some research concerning secure detection has been studied in a game-theoretic manner. For example, in [138], Vamvoudakis et al. consider the problem of estimating a binary random variable based on sensor measurements that may have been corrupted by an attacker. The estimation problem is formulated as a zero-sum partial information game. Then game-theoretic approaches are applied to derive the optimal detector. However, as that in [97], they focus on one-step scenario, while the strategy of each player in this chapter consists of an infinite sequence of behaviors. Note also that the binary sensor model in [138] restricts its application, and the explicit equilibrium is only obtained under certain conditions.

3.3 Problem Formulation

We consider the problem of detecting an unknown binary state $\theta \in \{0, 1\}$ with m sensors' measurements. At each time k, the measurement vector $y(k)$ is defined as:

$$y(k) \triangleq [y_1(k) \quad y_2(k) \quad \dots \quad y_m(k)] \in \mathbb{R}^m, \tag{3.1}$$

where $y_i(k)$ is the scalar measurement from sensor i at time k. The following assumptions on sensor measurement $y_i(k)$ are made:

1) Given θ, all measurements $\{y_i(k)\}_{i=1,\dots,m,k=1,\dots}$ are independent and identically distributed (i.i.d.).

2) For any Borel-measurable set $S \subseteq \mathbb{R}$, the probability of $y_i(k)$ belongs to S satisfies the following equation:

$$\mathbb{P}(y_i(k) \in S) = \begin{cases} \nu(S) & \text{if } \theta = 0 \\ \mu(S) & \text{if } \theta = 1 \end{cases}, \tag{3.2}$$

where μ and ν are the probability measure on $\mathbb{R}$. We further assume that ν and μ are absolutely continuous with respect to each other. Hence, the log-likelihood ratio $\lambda : \mathbb{R} \to \mathbb{R}$ of $y_i(k)$ is well defined as

$$\lambda(y_i(k)) \triangleq \log\left(\frac{d\mu}{d\nu}(y_i(k))\right), \tag{3.3}$$

where $d\mu/d\nu$ is the Radon-Nikodym derivative.

We denote by $Y(k)$ as the row vector of all measurements from time 1 to time k:

$$Y(k) \triangleq [y(1) \quad y(2) \quad \dots \quad y(k)] \in \mathbb{R}^{mk}. \tag{3.4}$$

At time k, define the detector $f_k : \mathbb{R}^{mk} \to [0,1]$ as a mapping from the measurement space $Y(k)$ to the interval $[0,1]$. The system follows the detection strategy like this: if $f_k(Y(k)) = \gamma \in [0,1]$, the system decides the detection value $\hat{\theta}$ to be 1 with probability γ, and decides $\hat{\theta}$ to be 0 with probability $1-\gamma$. The system's strategy $f \triangleq (f_1, f_2, \dots)$ is defined as an infinite sequence of detectors from time 1 to time infinity.

3.3.1 Attack Model

An attacker intends to disturb the detection state of the system by modifying sensors' measurements. However, because of the limited resource, it can only compromise n out of m sensors in the system. The set of the compromised sensors is denoted as $\mathcal{I} = \{i_1, \dots, i_n\}$, which is fixed over time. We assume that the system knows the number n, but it does not know the exact set $\mathcal{I}$.

To simplify notations, let us define:

$$y_{\mathcal{I}}(k) \triangleq [y_{i_1}(k) \quad y_{i_2}(k) \quad \dots \quad y_{i_n}(k)] \in \mathbb{R}^n, \tag{3.5}$$

and

$$Y_{\mathcal{I}}(k) \triangleq [y_{\mathcal{I}}(1) \quad y_{\mathcal{I}}(2) \quad \dots \quad y_{\mathcal{I}}(k)] \in \mathbb{R}^{nk}. \tag{3.6}$$

Now we consider the knowledge of the attacker. We assume that the attacker knows the probability measure ν and μ, the total number of sensors m, as well as the true state θ. We further characterize the attacker by its knowledge of the measurement vector:

1) An attacker is called a *weak* attacker if at any time k, it knows the measurement vector $Y_{\mathcal{I}}(k)$ from the compromised sensors;

2) An attacker is called a *strong* attacker if at any time k, it knows the measurement vector $Y(k)$ from all sensors.

Remark 3.1 *In practice, if the channel between the detector and sensors is not encrypted, then the attacker could potentially learn by eavesdropping on the measurements $Y(k)$ from all sensors and thus is a strong attacker. On the other hand, if the communication channel is encrypted and the attacker cannot listen to the communication between the uncompromised sensors and detector, then it is more suitable to assume a weak attacker model.*

For simplicity, let us denote by $\tilde{Y}(k)$ as the measurement vector known by the attacker at time k. From the above definition, we have

$$\tilde{Y}(k) \triangleq \begin{cases} Y_{\mathcal{I}}(k) & \text{for a weak attacker} \\ Y(k) & \text{for a strong attacker} \end{cases}.$$

At each time k, the attacker adds a *random* bias vector $y^a(k)$ according to its knowledge of the system $\tilde{Y}(k)$ to the true measurement $y(k)$. As a result, the system has to make its decision based on the manipulated measurement $y'(k)$ which can be defined as

$$y'(k) = y(k) + y^a(k) \triangleq [y'_1(k) \quad y'_2(k) \quad \ldots \quad y'_m(k)], \tag{3.7}$$

where $y'_i(k)$ is the manipulated measurement of sensor i at time k. Similar to (3.3), we define the log-likelihood ratio of $y'_i(k)$ as follows:

$$\lambda(y'_i(k)) \triangleq \log\left(\frac{d\mu}{d\nu}(y'_i(k))\right). \tag{3.8}$$

We further define

$$y^a(k) = [y^a_1(k) \quad y^a_2(k) \quad \ldots \quad y^a_m(k)] \triangleq g(\mathcal{I}, \theta, k, \tilde{Y}(k)), \tag{3.9}$$

where $y^a_i(k)_{i=1,\ldots,m}$ is the bias measurement vector added to sensor i at time k, and $y^a_i(k) = 0$ for $i \notin \mathcal{I}$. Obviously, g is a function of $\mathcal{I}, \theta, \tilde{Y}(k)$ and k. As a result, g characterizes the attacker's action for all possible scenarios. Hence, we can use g to denote the attacker's strategy. Similar to the definition of $Y(k)$, we further define the manipulated measurements from time 1 to k to be:

$$Y'(k) = [y'(1) \quad y'(2) \quad \ldots \quad y'(k)] \in \mathbb{R}^{mk}. \tag{3.10}$$

3.3.2 Asymptotic Detection Performance

Under attacks, the probability that the system makes a wrong decision at time k is

$$e(\theta, \mathcal{I}, k) \triangleq \begin{cases} \mathbb{E}f_k(Y'(k)) & \text{when } \theta = 0 \\ 1 - \mathbb{E}f_k(Y'(k)) & \text{when } \theta = 1 \end{cases}. \tag{3.11}$$

In this chapter, we are concerned with the worst-case scenario. To this end, let us define

$$\epsilon(k) \triangleq \max_{\theta=0,1,|\mathcal{I}|=n} e(\theta, \mathcal{I}, k), \tag{3.12}$$

which denotes the worst-case probability of detection error considering all possible sets of compromised sensors and true state θ.

At each time k, we want to design a system strategy f_k to minimize $\epsilon(k)$. However, since the computation of expectation usually involves complicated integration, we consider the *asymptotic detection performance* instead. Define the rate function as

$$\rho \triangleq \liminf_{k\to\infty} -\frac{\log \epsilon(k)}{k}. \tag{3.13}$$

Remark 3.2 *ρ indicates the rate that the probability of detection error goes to 0, which represents the detection performance of the system. From the definition (3.11)–(3.13), one can prove that ρ is always non-negative. If $\rho > 0$, then the probability of error will exponentially decay to 0, and a larger ρ indicates a shorter time for this convergence.*

From (3.11), it is trivial to know that the worst-case rate ρ is a function of both detection strategy f and attacker's strategy g. Therefore, in the rest of this chapter, we will use $\rho(f, g)$ instead to indicate this relationship.

3.3.3 Optimal Detection Rate for a Single Sensor in the Absence of Attacker

To simplify the presentation of the detection and attack strategies proposed later, in this subsection, we present the best rate can be achieved when only one sensor's measurements are used under the condition that the attacker is absent. We use C to denote this optimal rate.

From [24], this optimal decay rate is given by

$$C \triangleq \sup_{0<t<1} -\log\left[\mathbb{E}(e^{t\lambda(y_i(k))}|\theta = 0)\right], \tag{3.14}$$

where $\lambda(y_i(k))$ is the log-likelihood ratio defined in (3.3).

3.3.4 Nash-Equilibrium Strategy Pair

From the former discussion, clearly, the detector wants to maximize $\rho(f, g)$ to decrease the detection error, while the attacker wants to minimize it to make the error larger. Thus, in this section, we formulate the problem as a game between the detector and adversary, and intend to propose a pair of strategy (f^*, g^*), such that for any strategies f and g, the following inequality holds:

$$\rho(f^*, g) \geq \rho(f^*, g^*) \geq \rho(f, g^*). \tag{3.15}$$

As a result, the pair of strategy (f^*, g^*) reaches a Nash-equilibrium [44]. In other words, if the detector implements f^*, then there is no incentive for the adversary to deviate from g^*, and vice versa.

Remark 3.3 *In this section, we only provide one pair of equilibrium strategies in each case we investigate. However, it is worth noticing that the equilibrium strategy pair satisfying* (3.15) *may not be unique.*

3.4 Equilibrium Strategies for $m > 2n$

We first investigate the case when no more than half of the sensors are compromised by the attacker.

Before going on, we introduce the function $s(y, i, j) : \mathbb{R}^m \times \mathbb{N} \times \mathbb{N}$, where $1 \le i \le j \le m$, which satisfies the following two conditions:

1) For any permutation matrix P, $s(Py^T, i, j) = s(y, i, j)$.
2) If $y_1 \le y_2 \le \cdots \le y_m$, $s(y, i, j) = \sum_{l=i}^{j} y_l$.

Remark 3.4 *The function $s(y, i, j)$ can be interpreted as the summation from the ith element in vector y to the jth one after sorting in the ascending order.*

From the definition of $s(y, i, j)$, we have the following proposition:

Proposition 3.1 *For $y,\ y' \in \mathbb{R}^m$, and $||y - y'||_0 \le n$, the following inequalities holds:*

1) If $j + n \le m$, then $s(y', i, j) \le s(y, i + n, j + n)$;
2) If $i - n \ge 1$, then $s(y', i, j) \ge s(y, i - n, j - n)$.

To simplify notation, let us further define

$$\min_{m-2n}(y) \triangleq s(y, 1, m - 2n), \tag{3.16}$$

$$\operatorname*{med}_{m-2n}(y) \triangleq s(y, n + 1, m - n), \tag{3.17}$$

$$\max_{m-2n}(y) \triangleq s(y, 2n + 1, m). \tag{3.18}$$

Then we have the following lemma:

Lemma 3.1 *For $y,\ y' \in \mathbb{R}^m$, and $||y - y'||_0 \le n$, the following inequalities hold:*

$$\min_{m-2n}(y) \le \operatorname*{med}_{m-2n}(y') \le \max_{m-2n}(y). \tag{3.19}$$

Proof *The proof of Lemma 3.1 can be immediately achieved from Proposition 1 by substituting $n + 1$ to i, and $m - n$ to j.*

We are now ready to prove the main theorems of this section. We first derive a detection strategy which achieves the detection rate $\rho \ge m - 2n$ against any possible attack. After that, we propose an attack strategy and further prove that the rate for any detector cannot exceed $m - 2n$ against this attack. Therefore, the Nash-equilibrium is established.

3.4.1 Optimal Detection Strategy

At each time k, consider the following detection strategy f_k^*:

1) Compute the sum of log-likelihood ratio from time 1 to time k for each sensor i:

$$\Lambda_i'(k) = \sum_{t=1}^{k} \lambda(y_i'(t)), \tag{3.20}$$

where $\lambda(y_i'(t))$ is the log-likelihood ratio defined in (3.8).

Denote

$$\Lambda'(k) \triangleq [\Lambda_1'(k) \quad \Lambda_2'(k) \quad \ldots \quad \Lambda_m'(k)]. \tag{3.21}$$

2) Compute $\text{med}_{m-2n}(\Lambda'(k))$, and compare it to 0 to generate $\hat{\theta}$ as follows:

$$\hat{\theta} = \begin{cases} 0 & \text{if } \underset{m-2n}{\text{med}}(\Lambda'(k)) < 0 \\ 1 & \text{if } \underset{m-2n}{\text{med}}(\Lambda'(k)) \geq 0 \end{cases}. \tag{3.22}$$

The system's strategy is defined as $f^* \triangleq (f_1^*, f_2^*, \ldots)$.

Remark 3.5 *If (3.20) is done in a recursive fashion, then the computational complexity incurred at each k is $O(m)$. The computational complexity for (3.22) is $O(m \log(m))$, which can be achieved by first sorting $\Lambda_i(k)$ in the ascending order and then summing the middle $m - 2n$ elements. Therefore, the total computational complexity at each time step k is $O(m \log(m))$.*

We now have the first theorem in this chapter:

Theorem 3.1 *For any attack strategy g, the following inequality holds:*

$$\rho(f^*, g) \geq (m - 2n)C.$$

Proof *Define*

$$\Lambda_i(k) = \sum_{t=1}^{k} \lambda(y_i(t)), \tag{3.23}$$

and

$$\Lambda(k) \triangleq [\Lambda_1(k) \quad \Lambda_2(k) \quad \ldots \quad \Lambda_m(k)], \tag{3.24}$$

where $\Lambda_i(k)$ is defined in (3.23). Since the attacker can only manipulate up to n sensors, $||\Lambda(k) - \Lambda'(k)||_0 \leq n$. From Lemma 3.1, we have

$$\underset{m-2n}{\min}(\Lambda(k)) \leq \underset{m-2n}{\text{med}}(\Lambda'(k)) \leq \underset{m-2n}{\max}(\Lambda(k)). \tag{3.25}$$

Consider the situation when the true state $\theta = 0$. Following the above strategy f^*, the system will make a wrong decision if $\text{med}_{m-2n}(\Lambda'(k)) \geq 0$. As a result,

$$\begin{aligned} e(\theta = 0, \mathcal{I}, k) &= \mathbb{P}_0(\underset{m-2n}{\text{med}}(\Lambda'(k)) \geq 0) \\ &\leq \mathbb{P}_0(\underset{m-2n}{\max}(\Lambda(k)) \geq 0), \end{aligned}$$

where the inequality comes from (3.25).

Notice that $\max_{m-2n}(\Lambda(k)) \geq 0$ *if and only if there exists an index set* $\mathcal{K}$ *with cardinality* $m-2n$, *i.e.,* $|\mathcal{K}| = m - 2n$ *such that*

$$\sum_{i\in\mathcal{K}} \Lambda_i(k) \geq 0.$$

As a result,

$$\begin{aligned}
e(\theta = 0, \mathcal{I}, k) &\leq \mathbb{P}_0\left(\bigcup_{|\mathcal{K}|=m-2n}\left\{\sum_{i\in\mathcal{K}} \Lambda_i(k) \geq 0\right\}\right) \\
&\leq \sum_{|\mathcal{K}|=m-2n} \mathbb{P}_0\left(\sum_{i\in\mathcal{K}} \Lambda_i(k) \geq 0\right) \\
&= \binom{m}{2n}\mathbb{P}_0\left(\sum_{i=1}^{m-2n} \Lambda_i(k) \geq 0\right),
\end{aligned}$$

where the last equality holds because of the symmetry between sensors.

By Cramer's theorem [124],

$$-\limsup_{k\to\infty} \frac{\log \mathbb{P}_0(\sum_{i=1}^{m-2n} \Lambda_i(k) \geq 0)}{k} = (m-2n)C.$$

Therefore,

$$-\limsup_{k\to\infty} \frac{\log e(\theta = 0, \mathcal{I}, k)}{k} \geq (m-2n)C. \tag{3.26}$$

Similarly, one can prove that

$$-\limsup_{k\to\infty} \frac{\log e(\theta = 1, \mathcal{I}, k)}{k} \geq (m-2n)C. \tag{3.27}$$

Combining the two inequalities (3.26) *and* (3.27), *we get the conclusion that*

$$\rho(f^*, g) \geq (m-2n)C.$$

3.4.2 Optimal Attack Strategy

We consider the attack strategy g^* which flips the distribution of the compromised sensor measurements. Formally it is defined as follows:

1) The attacker generates i.i.d. random variables $y_i'(k)$, where $i = 1, \ldots, m$ and $k = 1, \ldots,$ such that the distribution of $y_i'(k)$ satisfies

$$\mathbb{P}(y_i'(k) \in S) = \begin{cases} \mu(S) & \text{if } \theta = 0 \\ \nu(S) & \text{if } \theta = 1 \end{cases}. \tag{3.28}$$

2) Compute $y_i^a(k)$ as follows:

$$y_i^a(k) = \begin{cases} y_i'(k) - y_i(k) & \text{if } i \in \mathcal{I} \\ 0 & \text{if } i \notin \mathcal{I} \end{cases}. \tag{3.29}$$

Theorem 3.2 *For any detection strategy f, the following inequality holds:*

$$\rho(f, g^*) \leq (m - 2n)C.$$

Proof *Consider the following two cases:*

1) True state $\theta = 0$ and sensor $1, 2, ..., n$ are compromised. In this case, at each time k, the sensor measurement $y(k)$ follows the following distribution:

$$y(k) \sim \underbrace{\mu \times \ldots \times \mu}_{n} \times \underbrace{\nu \times \ldots \times \nu}_{n} \times \underbrace{\nu \times \ldots \times \nu}_{m-2n}.$$

2) True state $\theta = 1$ and sensor $n+1, n+2, ..., 2n$ are compromised. In this case, at each time k, the sensor measurement $y(k)$ follows the following distribution:

$$y(k) \sim \underbrace{\mu \times \ldots \times \mu}_{n} \times \underbrace{\nu \times \ldots \times \nu}_{n} \times \underbrace{\mu \times \ldots \times \mu}_{m-2n}.$$

We use the probability measure μ_a and ν_a to denote the distribution of $y(k)$ in above two cases, respectively. Notice that for both cases, sensor 1 to sensor n will follow the distribution μ, and sensor $n+1$ to sensor $2n$ will follow the distribution ν.

Now we consider the following optimization problem which intends to minimize the probability of error in the above two cases:

$$\min \; \mathbb{P}_{\mu_a}(\hat{\theta} = 1) + \mathbb{P}_{\nu_a}(\hat{\theta} = 0), \tag{3.30}$$

where the first term indicates the probability of detection error in the first case, and the second term denotes this probability in the second case.

It is well-known optimal solution for (3.30) is the Bayes detector which is defined as follows [10]:

$$f_{\mathrm{B}}(Y'(k)) = \begin{cases} 0 & \text{if } \sum_{i=2n+1}^{m} \Lambda_i'(k) < 0 \\ 1 & \text{if } \sum_{i=2n+1}^{m} \Lambda_i'(k) \geq 0 \end{cases}.$$

Furthermore,

$$\begin{aligned} &\liminf_{k\to\infty} \frac{\log(\mathbb{P}_{\mu_a}(\hat{\theta} = 1) + \mathbb{P}_{\nu_a}(\hat{\theta} = 0))}{k} \\ &= \liminf_{k\to\infty} \frac{\log(e(\theta = 0, \mathcal{I}, k) + e(\theta = 1, \mathcal{I}, k))}{k} \\ &= \liminf_{k\to\infty} \log \frac{\log(\max_\theta(e(\theta, \mathcal{I}, k)))}{k}. \end{aligned}$$

As a result, Bayes detector is also optimal in the sense that the rate $\rho(f, g^*)$ is maximized. Notice that, this optimal detector only relies on the measurements from sensor $2n+1$ to sensor m for its decision. From Cramer's theorem [124],

$$-\limsup_{k\to\infty} \frac{\log \mathbb{P}_0(\sum_{i=2n+1}^{m} \Lambda_i'(k) \geq 0)}{k} = (m-2n)C,$$

and

$$-\limsup_{k\to\infty} \frac{\log \mathbb{P}_1(\sum_{i=2n+1}^{m} \Lambda_i'(k) < 0)}{k} = (m-2n)C,$$

Therefore, Bayes detector will distinguish the above two cases with the rate $(m-2n)C$. Because of its optimality, no detector can distinguish the above two cases with better than this rate against g^*. In other words, for any detection strategy f,

$$\rho(f, g^*) \leq (m-2n)C.$$

One can further prove that under such attacks, the best rate $(m-2n)C$ can also be achieved by the optimal detection strategy f^* defined in (3.20)-(3.22). As a result, from Theorem 3.1 and Theorem 3.2, we can immediately derive the following theorem:

Theorem 3.3 *The strategy pair (f^*, g^*) forms a Nash-equilibrium such that*

$$\rho(f, g^*) \leq \rho(f^*, g^*) \leq \rho(f^*, g),$$

where f^* is the optimal detection strategy defined in (3.20)–(3.22), g^* is the optimal attack strategy defined in (3.28)–(3.29), and $\rho(f^*, g^*) = (m-2n)C$.

3.5 Equilibrium Strategies for $m \leq 2n$

In this section, we consider the case when more than half of the sensors are compromised.

We begin with the attack strategy g^* defined as below:

1) The attacker generates i.i.d. random variables $y_i'(k)$, where $i = 1, \ldots, m$ and $k = 1, \ldots,$ such that

$$\mathbb{P}(y_i'(k) \in S) = \begin{cases} \mu(S) & \text{if } \theta = 0 \\ \nu(S) & \text{if } \theta = 1 \end{cases}. \tag{3.31}$$

2) Compute $y_i^a(k)$ as follows:
If $\theta = 0$,

$$y_i^a(k) = \begin{cases} y_i'(k) - y_i(k) & \text{if } i \in \mathcal{J}_1 \\ 0 & \text{if } i \notin \mathcal{J}_1 \end{cases}; \tag{3.32}$$

If $\theta = 1$,

$$y_i^a(k) = \begin{cases} y_i'(k) - y_i(k) & \text{if } i \in \mathcal{J}_2 \\ 0 & \text{if } i \notin \mathcal{J}_2 \end{cases}, \tag{3.33}$$

where $\mathcal{J}_1, \mathcal{J}_2$ are the subsets of the compromised sensors set when $\theta = 0$ and $\theta = 1$, respectively, with $|\mathcal{J}_1| = \lceil \frac{m}{2} \rceil$, and $|\mathcal{J}_2| = \lfloor \frac{m}{2} \rfloor$.

In other words, under g^*, the attacker will flip the measurements' distribution of sensors in set $\mathcal{J}_1, \mathcal{J}_2$, when $\theta = 0$ and 1, respectively.

Remark 3.6 *The reason why the adversary will not implement the same strategy as* (3.28)–(3.29) *in the situation $m \leq 2n$ is that under such attacks, the detector can easily figure it out by simply flipping the compromised sensors' measurements back if it knows the strategy of the adversary. Thus, the detection rate ρ would not be minimized.*

Theorem 3.4 *For any detection strategy f, the following inequality holds:*

$$\rho(f, g^*) = 0.$$

Proof *Consider the following two cases:*

1) True state $\theta = 0$, $\mathcal{I} = \{1, \ldots, n\}$, and $\mathcal{J}_1 = \{1, \ldots, \lceil m/2 \rceil\}$, then the distribution of the sensor measurement $y(k)$ at each time k is as follows:

$$y(k) \sim \underbrace{\mu \times \mu \times \ldots \times \mu}_{\lceil \frac{m}{2} \rceil} \times \underbrace{\nu \times \nu \times \ldots \times \nu}_{m - \lceil \frac{m}{2} \rceil}.$$

2) True state $\theta = 1$, $\mathcal{I} = \{m - n + 1, \ldots, m\}$, and $\mathcal{J}_2 = \{\lceil m/2 \rceil + 1, \ldots, m\}$, then the distribution of the sensor measurement $y(k)$ at each time k is as follows:

$$y(k) \sim \underbrace{\mu \times \mu \times \ldots \times \mu}_{\lceil \frac{m}{2} \rceil} \times \underbrace{\nu \times \nu \times \ldots \times \nu}_{m - \lceil \frac{m}{2} \rceil}.$$

Since the distribution of $y(k)$ is identical, no detector can distinguish the above two cases. Therefore, Theorem 3.4 follows immediately.

From Theorem 3.4, we have the next theorem:

Theorem 3.5 *For any detection strategy f, the strategy pair (f, g^*) forms a Nash-equilibrium such that*

$$\rho(f, g^*) = 0 \leq \rho(f, g),$$

where g^* is the attack strategy defined in (3.31)–(3.33).

Proof *The proof of Theorem 3.5 is obvious since ρ is always nonnegative.*

Remark 3.7 *Since g^* in* (3.28)–(3.29) *and* (3.31)–(3.33) *only requires the attacker's knowledge of the compromised sensors' measurements. Hence, equilibrium strategy pair in Theorems 3.3 and 3.5 can be achieved by even the weak attacker.*

3.6 Extension

In practice, the attacker may not be present consistently. Thus, the system with all sensors uncompromised may operate for some time. Usually, the performance of the detection rule when there is no attacker at all is referred to by efficiency, while the performance when the attacker is present is referred to by security. In this section, we investigate the efficiency of our proposed detection strategy.

Theorem 3.6 *Under the detection rule (3.20)–(3.22), when all the sensors are benign, the detector will achieve the detection rate of* $(m-n)C$.

Proof *Since there is no attacker in this situation, we will use* $\rho(f^*)$ *rather than* $\rho(f^*, g)$ *to denote the detection rate.*

Consider the situation when $\theta = 0$. *Notice that*

$$e(\theta = 0, \mathcal{I} = \emptyset, k) = \mathbb{P}_0(s(\Lambda(k), n+1, m-n) \geq 0).$$

Hence, we are interested in the probability of the event $\{s(\Lambda(k), n+1, m-n) \geq 0\}$.

Notice that

$$\begin{aligned} e(\theta = 0, \mathcal{I} = \emptyset, k) &= \mathbb{P}_0(s(\Lambda(k), n+1, m-n) \geq 0) \\ &\leq \mathbb{P}_0(s(\Lambda(k), n+1, m) \geq 0) \\ &\leq \binom{m}{n} \mathbb{P}_0\left(\sum_{i=1}^{m-n} \Lambda_i(k) \geq 0\right). \end{aligned}$$

Then, from Cramer's theorem [124],

$$-\limsup_{k\to\infty} \frac{\log e(\theta = 0, \mathcal{I} = \emptyset, k)}{k} \geq (m-n)C.$$

On the other hand, we assume $\mathbb{P}_0(\Lambda_i(k) \geq 0) = M$, *and from [124],*

$$-\limsup_{k\to\infty} \frac{\log M}{k} = C. \tag{3.34}$$

If $\Lambda_1(k) < 0, \ldots, \Lambda_n(k) < 0$, *and* $\Lambda_{n+1}(k) \geq 0, \ldots, \Lambda_m(k) \geq 0$, *then the considered event* $\{s(\Lambda(k), n+1, m-n) \geq 0\}$ *will happen. Therefore, the probability of it is lower bounded by* $\binom{m}{n} M^{m-n}(1-M)^n$. *As a result,*

$$\begin{aligned} &-\limsup_{k\to\infty} \frac{\log e(\theta = 0, \mathcal{I} = \emptyset, k)}{k} \\ &\leq -\limsup_{k\to\infty} \frac{\log\left(\binom{m}{n} M^{m-n}(1-M)^n\right)}{k} \\ &= (m-n)C. \end{aligned}$$

Similar results can be achieved under the condition $\theta = 1$. As a result,

$$\rho(f^*) = (m - n)C.$$

Remark 3.8 *We notice that the rate in Theorem 3.6 is not optimal because $\rho(f)$ is not maximized, since one can prove that if the attacker is absent, then the Bayes detector [10] will achieve the best rate of mC. This means that in order to increase the security, we sacrifice the system's efficiency to some degree. One can further derive that the larger the n is, the more resilient the detector will be under attacks, but at the same time, the more performance degradation will occur during normal operation when the attacker is absent. Therefore, there exists a trade-off between security and efficiency, which can be tuned by choosing a suitable parameter n.*

3.7 Numerical Example

In this part, we provide numerical examples to verify the theoretical results established in the previous subsections. We assume that the sensor's measurement $\{y_i(k)\}_{i=1,\ldots,m,k=1,\ldots}$ follows the distribution $\mathcal{N}(-1,1)$[1] when $\theta = 0$, and follows $\mathcal{N}(1,1)$ when $\theta = 1$. From (3.14), one can derive that the optimal decay rate of a single sensor is $C = 0.5$.

Since the situation when $m \leq 2n$ is trivial, we only focus on the case where $m > 2n$. We first assume $m = 7$, and n varies from 0 to 3. Figure 3.1 shows that under the detection strategy f^* defined in (3.20)-(3.22) and attack strategy g^* defined in (3.28)-(3.29), the detection rate $\rho(f^*, g^*)$ finally approaches $0.5(m - 2n)$, i.e., $(m - 2n)C$, which is consistent with our result.

3.8 Conclusion

This chapter studies the binary hypothesis testing in an adversarial environment. The detector determines the true state of an unknown parameter based on the measurements from m sensors, out of which n sensors might be arbitrarily compromised. The exponent rate, at which the worst-case probability of detection error goes to 0, is adopted as the performance metric. Obviously, the attacker intends to deteriorate the detection performance by maximizing this rate, while the detector wants to minimize it. This problem is thus formulated as a game between these two players. We study both cases where

[1] $\mathcal{N}(a, b)$ represents the Gaussian distribution with mean equals to a, and variance equals to b.

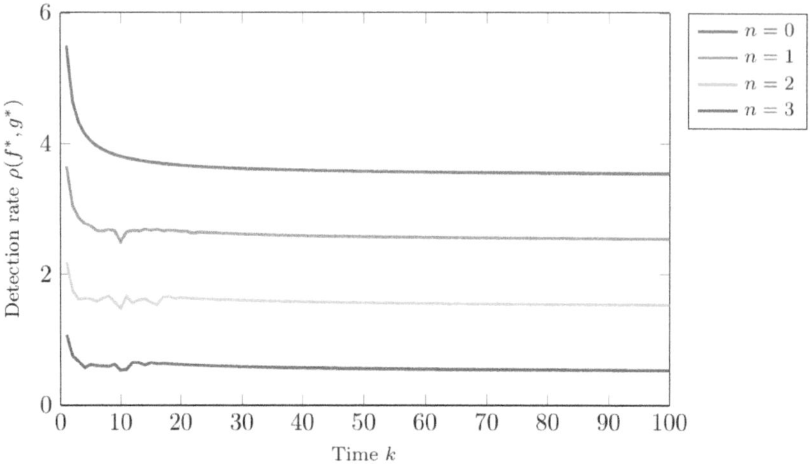

FIGURE 3.1: Detection rate when $m = 7$ and $m > 2n$ under the proposed optimal strategy pair.[2]

$m > 2n$ and $m \leq 2n$, and obtain an equilibrium strategy pair for the players under both cases.

[2]Reproduced with permission of ©2013 IEEE.

4

Resilient Consensus of Second-Order Systems through Impulsive Control

4.1 Introduction

Recent advances in signal processing and cooperative control have led to growing research interests in multi-agent systems (MASs). One of the most important focuses in MASs is the canonical consensus. Given a set of autonomous agents, it seeks an agreement upon certain quantities of interest and has wide applications in biological, social, and engineering worlds, like animal groups, sensor networks, and robotic teams [107].

Considerable attention has been paid to the development of consensus algorithms in MASs with first-order dynamics; see [106, 115, 142] for examples. Meanwhile, inspired by many real-world applications, there are also growing interests in second-order consensus algorithms, where the agents are governed by both position and velocity states. The insights into the second-order consensus problems also shed light on introducing more realistic dynamics into the individual agent's model based on the general framework of MASs and are especially meaningful for the implementation of cooperative control strategies in engineering networked systems [157].

In these years, various algorithms have been proposed to achieve the second-order consensus (see [113, 158, 51]). Such protocols are normally based on the hypothesis that every computing agent is trustworthy and cooperate to follow the algorithms throughout their execution. Nevertheless, as the scale of the network increases, it becomes more difficult to secure every agent. On one hand, autonomous agents will communicate with each other to make control decisions. This opens the system to malicious attacks. On the other hand, some agents may not be willing to follow the given rules if they weigh their private interests more than the public ones. It is reported that most of the existing algorithms are fragile to such network misbehaviors, which can prevent the network from reaching an agreement ([72, 32, 130]). Since the consensus algorithms have been widely applied in safety-critical systems, and serve as the basis of distributed computing and control, the studies on *resilient consen-*

DOI: 10.1201/9781003409199-4

sus have gained a growing research attention. Particularly, it aims to design distributed protocols to guarantee the agreement among non-faulty ones.

In this chapter, we investigate the resilient consensus in continuous-time second-order MASs, where some nodes might be faulty or adversarial. Despite such malicious and unexpected behaviors, the normal agents still aim to reach an agreement among each other. To further avoid the continuous communication among nodes, a resilient impulsive algorithm is proposed, where the signal transmission and control action are allowed at the aperiodic sampling instants. At each sampling time, the normal agent removes the most extreme values in the neighborhood and derives its control signal with the remaining ones. Sufficient conditions related to the network topology and tolerable number of misbehaving nodes are established to achieve the resilient consensus while reducing the communication cost.

4.2 Related Work

The resilient consensus of first-order systems has been studied in the literature over years. Most approaches adopt the idea of simply ignoring the suspicious values. For instance, Dolev et al. consider the approximate resilient agreement in a complete network [38] with some of nodes being faulty. In order to overrule the malicious effects from misbehaving agents, a Mean-Subsequence Reduced (MSR) algorithm is developed, where each normal node discards the most extreme values from neighbors and makes updates with the average of the remaining ones. This protocol has then inspired a series of protocols (see [64, 137]). In a more recent work [72], the authors present a modified version of MSR, named as Weighted-MSR. Compared with that in MSR, the normal node only excludes the most suspicious ones that are strictly larger or smaller than its own. Different from [38], the exact consensus can be guaranteed. Moreover, instead of the complete graph, the resilience of Weighted-MSR is characterised in terms of network robustness. These strategies are proved to secure the normal agents from being seriously affected by the misbehaviors, and thus ensure the (approximate) resilient agreement in hostile environment.

Dibaji et al. recently generalize the above results to the second-order systems [32, 34]. To facilitate the agreement, equidistant sampling intervals are required to discretize the system and synthesize the controller. This assumption yet prevents their results from being applied to systems with time-varying or even uncertain sampling period. Moreover, the velocity information is required in their works for the design of control signal. As a contrast, in this chapter, we study the resilient consensus in the network with uncertain sampling interval and using only position information. Adopting the common idea of most resilient protocols, each healthy agent ignores the most extreme states in its updates. The main contributions are summarized as below:

1. Despite the fact that the second-order consensus has been widely investigated in the literature for benign environment, the unpredictable behavior of misbehaving nodes complicates the convergence analysis and prevents the methods therein from being directly applied. To overcome it, we present an equivalent model of the closed-loop system, by integrating the attack model with resilient controller.

2. The coupling between position and velocity states also makes it difficult to directly use the analysis methods of first-order resilient algorithms, whose efficiency is proved through convex analysis. We therefore resort to *Sarymsakov matrices* to exploit key features of the equivalent model. Sufficient topological conditions are finally derived to ensure the resilient consensus.

4.3 Problem Formulation

Let us consider a second-order system with n agents, who cooperate over the network $\mathcal{G} = \{\mathcal{V}, \mathcal{E}\}$. The dynamics of agent $i \in \mathcal{V}$, is described by

$$\begin{aligned} \dot{x}_i(t) &= v_i(t), \\ \dot{v}_i(t) &= u_i(t), \end{aligned} \tag{4.1}$$

where $u_i(t)$ denotes the control input, and $x_i(t), v_i(t) \in \mathbb{R}$ are respectively the position and velocity states of agent i. In this chapter, we consider the scenario where the communication among agents possibly occur only at sampling instants. The sequence of sampling time $\{t_k\}$ satisfies that $0 = t_0 < t_1 < \ldots < t_k < \ldots$ and $\lim_{k\to\infty} t_k = \infty$.

Let $h_k = t_k - t_{k-1}$ be the sampling period between t_k and t_{k-1}. We further assume that $\underline{h} \leq h_k \leq \overline{h}, \forall k \in \mathbb{Z}$, where $\underline{h}$ and $\overline{h}$ are respectively the lower and upper bounds for sampling intervals.

Remark 4.1 *For simplicity and clear presentation of our ideas, this part considers the system with scalar states. However, the entire analysis can be readily extended to multi-dimensional scenarios, by applying the developed algorithm to each entry of the vector states.*

Remark 4.2 *Compared with the existing work with sampled control (e.g. [80, 158]), the sampling period in this work is not necessary to be constant and thus allowed to be aperiodic.*

Consensus in the multi-agent network (4.1) is said to be achieved if the following hold for any initial states and any $i, j \in \mathcal{V}$:

$$\begin{aligned} \lim_{t\to\infty} |x_i(t) - x_j(t)| &= 0, \\ \lim_{t\to\infty} |v_i(t)| &= 0. \end{aligned} \tag{4.2}$$

4.3.1 Attack Model

In this chapter, an adversarial environment is discussed, where some of the agents might be faulty or misbehaving. For simplicity, let us denote the set of such agents as $\mathcal{A}$. Any agent $i \in \mathcal{A}$ is the one that either manipulated by the attacker, or failing to follow the pre-defined update rule. On the other hand, the normal or benign agents will always obey the control strategy, whose set is denoted as $\mathcal{R}$. Without loss of generality, let $\mathcal{R} = \{1, 2, ..., B\}$, where $B \triangleq |\mathcal{R}|$. Clearly, $\mathcal{R} \cap \mathcal{A} = \varnothing$ and $\mathcal{R} \cup \mathcal{A} = \mathcal{V}$.

Given the limited energy of adversaries, it is reasonable to assume an upper bound on the number of misbehaving nodes. In specific, the network misbehaviors could be characterized by the following manner:

Definition 4.1 *The network is said to under an f-local attack if for any normal agent, no more than f misbehaving ones exist in its neighborhood, i.e., $|\mathcal{A} \cap \mathcal{N}_i| \leq f, \forall i \in \mathcal{B}$.*

Note the considered f-local attack model is first introduced in [66], and has been widely adopted in the study of networked systems since then (see [69, 38, 72, 130] for examples). In this thesis, we focus on the worst-case situation, where no restrictions are imposed on the transmitted information of agent $i \in \mathcal{A}$. Specifically, the misbehaving agents are allowed to send arbitrary and different data to different neighbors. They could even collude with each other to decide on the false values to be transmitted.

The network misbehaviors would greatly jeopardize the performance of standard consensus algorithms, e.g., [84]. As one might imagine, if no security strategies equipped, even a single misbehaving node could be able to control the evolution of normal states on its desire.

4.3.2 Resilient Consensus

The above security concerns necessitate the design of resilient consensus protocol, aiming to achieve the below objective:

Resilient consensus: The multi-agent network (4.1) achieves resilient consensus, if (4.2) holds for any initial states and any $i, j \in \mathcal{R}$, regardless of the misbehaviors.

Hence through resilient protocols, the network could be avoided of being seriously affected by the network misbehaviors. To proceed, the following assumption is also imposed regarding $\mathcal{G} = (\mathcal{V}, \mathcal{E})$:

Assumption 4.1 *For any normal agent, it has at least $2f + 1$ neighbors.*

4.4 Resilient Impulsive Algorithms

This section provides a resilient consensus protocol. To avoid the continuous information exchange among agents, for each agent an impulsive control signal is designed to occur at only sampling instants. Furthermore, given that the velocity states of agents are often unavailable in practical applications, we involve the controller with only position states.

While the misbehaving agents can perform arbitrarily, the normal ones should always follow the designed protocol. More specifically, each normal agent $i \in \mathcal{R}$ updates as outlined in Algorithm 5.1:

Algorithm 4.1 Resilient impulsive consensus protocol

At any time $t \in \mathbb{R}_{>0}$ **do**
if $t = t_k$ **then**

1) Collect the values $x_j(t_k), j \in \mathcal{N}_i$ in $\mathcal{X}_i(t_k)$.

2) In $\mathcal{X}_i(t_k)$, remove f largest values that are higher than agent i's own state $x_i(t_k)$. If there are less than f values higher than $x_i(t_k)$, then remove all of them. Similarly, remove f smallest values that are less than $x_i(t_k)$. If there are less than f values lower than $x_i(t_k)$, then remove all these values.

3) Denote $\mathcal{J}_i(t_k)$ as the set of agents whose values are retained after 2). Design $u_i(t_k)$ as

$$\begin{aligned} u_i(t_k) = \Big[\xi_1 \sum_{j\in\mathcal{J}_i(t_k)} & a_{ij}(t_k)(x_j(t_k) - x_i(t_k)) \\ & - \xi_2(x_i(t_k) - x_i(t_{k-1}))\Big]\delta(t-t_k), \end{aligned} \tag{4.3}$$

where $\delta(\cdot)$ is the Dirac impulsive function, $a_{ij}(t_k) > 0$ and $\sum_{j\in\mathcal{J}_i(t_k)} a_{ij}(t_k) < 1$, ξ_1, ξ_2 are design parameters chosen according to the conditions in (4.12) given later.

else

$u_i(t) = 0.$

end if

In view of the proposed strategy, it adopts a similar idea to that of many resilient algorithms, namely to discard the most suspicious values. In what follows, we shall theoretically prove the efficiency of such a protocol.

4.5 Convergence Analysis

To evaluate the performance of the proposed resilient algorithm, we first present the following lemma:

Lemma 4.1 *Consider the digraph $\mathcal{G} = (\mathcal{V}, \mathcal{E})$. Suppose the network is under f-local attack and the normal agent i updates with the proposed strategy. Under Assumption 4.1, there exists a nonempty set $\mathcal{M}_i(t_k) \subset \mathcal{N}_i \cap \mathcal{R}$ and weights $\{\bar{a}_{ij}(t_k)\}$, such that the control input (4.3) is mathematically equivalent to*

$$
\begin{aligned}
u_i(t_k) = \Big[\xi_1 \sum_{j \in \mathcal{M}_i(t_k)} \bar{a}_{ij}(t_k)(x_j(t_k) - x_i(t_k)) \\
- \xi_2(x_i(t_k) - x_i(t_{k-1}))\Big]\delta(t - t_k),
\end{aligned}
\tag{4.4}
$$

with the weights satisfying that $\sum_{j \in \mathcal{M}_i(t_k)} \bar{a}_{ij}(t_k) < 1$.

Proof *We prove the result by construction. Note that at most $2f$ values in $\mathcal{X}_i(t_k)$ would be discarded. Therefore, under Assumption 4.1, one obtains that $\mathcal{J}_i(t_k) \neq \varnothing$. Then let us consider the following cases:*

- *$\mathcal{J}_i(k) \cap \mathcal{A} = \varnothing$, i.e., there is no adversarial agent in $\mathcal{J}_i(k)$. In this scenario, the construction of $\bar{a}_{ij}(t_k)$ is trivial by simply making it equal $a_{ij}(t_k)$.*
- *$\mathcal{J}_i(k) \cap \mathcal{A} \neq \varnothing$. Now suppose some misbehaving agents exist in $\mathcal{J}_i(t_k)$. Consider any agent $j \in \mathcal{J}_i(t_k) \cap \mathcal{A}$. Since $x_j(t_k)$ is retained by agent i, it must hold that either there are f values in $\mathcal{X}_i(t_k)$ no less than $x_j(t_k)$, or agent i's own value $x_i(t_k)$ is not less than $x_j(t_k)$. Similarly, it is also true that either f neighboring values are not greater than $x_j(t_k)$, or $x_i(t_k) \leq x_j(t_k)$. As no more than f faulty ones exist in agent i's neighborhood, one could always find a pair of normal agents $p, q \in \mathcal{N}_i \cup \{i\}$, such that $x_p(t_k) \leq x_j(t_k) \leq x_q(t_k)$. Hence, we have $x_j(t_k) = \gamma x_p(t_k) + (1-\gamma)x_q(t_k)$ for some $0 \leq \gamma \leq 1$. By setting $\bar{a}_{ip}(t_k) = a_{ip}(t_k) + \gamma a_{ij}(t_k)$ and $\bar{a}_{iq}(t_k) = a_{iq}(t_k) + (1-\gamma)a_{ij}(t_k)$, the contribution of any misbehaving node j can be transformed to that of two normal ones (i.e., agents p and q). Moreover, $\bar{a}_{ip}(t_k)+\bar{a}_{iq}(t_k) = a_{ip}(t_k)+a_{ij}(t_k)+a_{iq}(t_k)$. By repeating the above analysis for each misbehaving agent in $\mathcal{J}_i(t_k)$, the conclusion can be derived and the proof completes.*

As indicated by Lemma 4.1, the evolution of any normal agent's state only relies on the benign ones in its neighborhood. Hence, the misbehaving agents are unable to have arbitrary control over these normal ones. As a result, the healthy agents are protected from being affected by the misbehaviors too much.

Based on (6.3), let us define $\bar{A}(t_k) \triangleq (\bar{a}_{ij}(t_k))$ with

$$
\bar{a}_{ii}(t_k) = 1 - \sum_{j \in \mathcal{M}_i(t_k)} \bar{a}_{ij}(t_k) > 0, \ \forall i \in \mathcal{R}. \tag{4.5}
$$

The below result will be helpful in our further analysis:

Lemma 4.2 *Consider the digraph $\mathcal{G} = (\mathcal{V}, \mathcal{E})$. Suppose the network is under f-local attack, satisfies Assumption 4.1 and is with $(2f+1)$-robustness. Let $\bar{A}(t_k)$ be the one established by (4.5) at time $t = t_k$. Then for each $k \in \mathbb{N}$, $\bar{A}(t_k) = \{\bar{a}_{ij}(t_k)\}$ is a Sarymsakov matrix.*

Proof *Consider any disjoint and nonempty subsets $\mathcal{R}_1, \mathcal{R}_2 \subsetneq \mathcal{V}$. Under Definition 2.2, there must exist an agent, denoted as agent i, in $\mathcal{R}_1 \cup \mathcal{R}_2$, such that it has at least $2f+1$ neighbors outsides. Without loss of generality, let $i \in \mathcal{R}_1$ who has more than $2f+1$ neighbors outside $\mathcal{R}_1$. Since no more than $2f$ values would be discarded, at least one of these $2f+1$ ones would be retained by agent i. As indicated by the proof of Lemma 4.1, this value must either be a normal agent's state or a convex combination of two normal agents' states. Hence there exists some normal node $j \in \mathcal{R} \backslash \mathcal{R}_1$ such that $j \in F_{\bar{A}(t_k)}(\mathcal{R}_1)$. Then one of the following cases must hold:*

1) $j \in F_{\bar{A}(t_k)}(\mathcal{R}_2)$, then $F_{\bar{A}(t_k)}(\mathcal{R}_1) \cap F_{\bar{A}(t_k)}(\mathcal{R}_2) \neq \varnothing$;

2) $j \notin F_{\bar{A}(t_k)}(\mathcal{R}_2)$ and $F_{\bar{A}(t_k)}(\mathcal{R}_1) \cap F_{\bar{A}(t_k)}(\mathcal{R}_2) = \varnothing$. Note that for any $l \in \mathcal{R}_1, \bar{a}_{ll}(t_k) > 0$ and thus $l \in F_{\bar{A}(t_k)}(\mathcal{R}_1)$. Similarly, $p \in F_{\bar{A}(t_k)}(\mathcal{R}_2)$, $\forall p \in \mathcal{R}_2$. Therefore, $(F_{\bar{A}(t_k)}(\mathcal{R}_1) \cup F_{\bar{A}(t_k)}(\mathcal{R}_2)) \supset (\mathcal{R}_1 \cup \mathcal{R}_2 \cup \{j\})$, where last three sets are disjoint. Hence, $|F_{\bar{A}(t_k)}(\mathcal{R}_1) \cup F_{\bar{A}(t_k)}(\mathcal{R}_2)| \geq |\mathcal{R}_1| + |\mathcal{R}_2| + 1 > |\mathcal{R}_1| + |\mathcal{R}_2| = |\mathcal{R}_1 \cup \mathcal{R}_2|$.

Recalling Definition 2.7, any $\bar{A}(k)$ is Sarymsakov.

In view of Lemma 4.1, the network with resilient impulsive control can be described by the following equations:

$$
\begin{aligned}
&\dot{x}_i(t) = v_i(t), \\
&\dot{v}_i(t) = 0, \quad t \in (t_k, t_{k+1}], \\
&\Delta v_i(t_k) = -\xi_1 \sum_{j \in \mathcal{R}} \bar{l}_{ij}(t_k) x_j(t_k) - \xi_2 (x_i(t_k) - x_i(t_{k-1})),
\end{aligned} \tag{4.6}
$$

where $\Delta v_i(t_k) = v_i(t_k^+) - v_i(t_k), v_i(t_{k+1}) = v_i(t_k^+) = \lim_{t \to t_k^+} v_i(t)$ and

$$
\bar{l}_{ij}(t_k) = \begin{cases} \sum_{j \in \mathcal{M}_i(t_k)} \bar{a}_{ij}(t_k), & j = i \\ -\bar{a}_{ij}(t_k), & j \in \mathcal{M}_i(t_k) \\ 0, & j \neq i \text{ and } j \notin \mathcal{M}_i(t_k) \end{cases}.
$$

Consider the close-loop system dynamics at sampling times:

$$
x_i(t_{k+1}) = x_i(t_k) + h_{k+1} v_i(t_k^+), \tag{4.7}
$$

and

$$\begin{aligned}
& v_i(t_{k+1}^+) \\
&= v_i(t_k^+) - \xi_1 \sum_{j\in\mathcal{R}} \bar{l}_{ij}\left(t_{k+1}\right) x_j\left(t_{k+1}\right) - \xi_2\left(x_i\left(t_{k+1}\right) - x_i(t_k)\right) \\
&= v_i(t_k^+) - \xi_1 \sum_{j\in\mathcal{R}} \bar{l}_{ij}\left(t_{k+1}\right) [x_j(t_k) + h_{k+1} v_j(t_k^+)] - \xi_2 h_{k+1} v_i(t_k^+) \\
&= (1-\xi_2 h_{k+1}) v_i(t_k^+) - \xi_1 \sum_{j\in\mathcal{R}} \bar{l}_{ij}\left(t_{k+1}\right) x_j(t_k) - \xi_1 h_{k+1} \sum_{j\in\mathcal{R}} \bar{l}_{ij}\left(t_{k+1}\right) v_j(t_k^+).
\end{aligned} \tag{4.8}$$

Inspired by [84], let us define auxiliary states as $\tilde{x}_i(k) = x_i(t_k)$ and $\tilde{v}_i(k) = x_i(t_k) + 2/\xi_2 v_i(t_k^+)$. From (4.7) and (4.8), one has

$$\begin{aligned}
\tilde{x}_i(k+1) &= \left(1 - \frac{\xi_2 h_{k+1}}{2}\right)\tilde{x}_i(k) + \frac{\xi_2 h_{k+1}}{2}\tilde{v}_i(k), \\
\tilde{v}_i(k+1) &= \frac{\xi_2 h_{k+1}}{2}\tilde{x}_i(k) + \left(1 - \frac{\xi_2 h_{k+1}}{2}\right)\tilde{v}_i(k) - (2/\xi_2 - h_{k+1})\xi_1 \sum_{j\in\mathcal{R}} \bar{l}_{ij}\left(t_{k+1}\right)\tilde{x}_j(k) \\
&\quad - \xi_1 h_{k+1} \sum_{j\in\mathcal{R}} \bar{l}_{ij}\left(t_{k+1}\right)\tilde{v}_j(k).
\end{aligned}$$

Now let $\tilde{x}(k) = [\tilde{x}_1(k), \ldots, \tilde{x}_B(k)]^T$, $\tilde{v}(k) = [\tilde{v}_1(k), \ldots, \tilde{v}_B(k)]^T$ and $y(k) = [\tilde{x}(k); \tilde{v}(k)]$. Then

$$y(k+1) = W(k+1)y(k), \tag{4.9}$$

where

$$W(k) = \begin{bmatrix} W_{11}(k) & W_{12}(k) \\ W_{21}(k) & W_{22}(k) \end{bmatrix}, \tag{4.10}$$

and

$$\begin{aligned}
W_{11}(k) &= \left(1 - \frac{\xi_2 h_k}{2}\right) I, \\
W_{12}(k) &= \frac{\xi_2 h_k}{2} I, \\
W_{21}(k) &= \frac{\xi_2 h_k}{2} I - \xi_1\left(\frac{2}{\xi_2} - h_k\right)\bar{L}(k), \\
W_{22}(k) &= \left(1 - \frac{\xi_2 h_k}{2}\right) I - \xi_1 h_k \bar{L}(k),
\end{aligned} \tag{4.11}$$

with $\bar{L}(k) = \{\bar{l}_{ij}(t_k)\}$ and $F(k) = [\mathbf{0}_{\mathcal{B}}; \bar{L}(k)]$.

Clearly, the network achieves resilient consensus, if and only if $\lim_{k\to\infty} y(k) = \beta \mathbf{1}_{2B}$ for some $\beta \in \mathbb{R}$. Therefore, it is sufficient to study the convergence of (4.9). To this end, we first characterise $W(k)$ in the following lemma:

Lemma 4.3 *If the parameters in (4.3) satisfy that*

$$\begin{aligned} 0 <& \xi_2 < 2/\overline{h}, \\ 0 <& \xi_1 < \min\left\{ \frac{(\xi_2 \underline{h})^2}{2\underline{h}(2-\xi_2\underline{h})}, \frac{2-\xi_2\overline{h}}{2\overline{h}} \right\}, \end{aligned} \tag{4.12}$$

then $W(k)$ is Sarymsakov at each step k.

Proof *We shall first show $W(k)$ is row stochastic at any k. Given $\mathbf{1}_B \bar{L}(k) = 0$, each row of $W(k)$ sums to 1. Under (4.12), the non-zero elements of $W_{11}(k)$ and $W_{12}(k)$ are strictly positive. Moreover, since $\xi_1 < \frac{(\xi_2\underline{h})^2}{2\underline{h}(2-\xi_2\underline{h})}$, one has*

$$\begin{aligned} &\frac{\xi_2 h_k}{2} - \xi_1\left(\frac{2}{\xi_2} - h_k\right)\bar{l}_{ii}(k) \\ >& \frac{\xi_2\underline{h}}{2} - \frac{(\xi_2\underline{h})^2}{2\underline{h}(2-\xi_2\underline{h})}\left(\frac{2}{\xi_2} - \underline{h}\right) = 0, \end{aligned} \tag{4.13}$$

where the first inequality is because $\bar{l}_{ii}(k) < 1, \forall i \in \mathcal{B}$. Similarly, we obtain that

$$\begin{aligned} \left(1 - \frac{\xi_2 h_k}{2}\right) - \xi_1 h_k \bar{l}_{ii}(k) >& \left(1 - \frac{\xi_2 \overline{h}}{2}\right) - \xi_1 \overline{h} \\ >& \left(1 - \frac{\xi_2 \overline{h}}{2}\right) - \frac{2-\xi_2\overline{h}}{2\overline{h}}\overline{h} = 0. \end{aligned} \tag{4.14}$$

Since each non-diagonal element in $W_{21}(k)$ and $W_{22}(k)$ is non-negative under (4.12), $W(k)$ is row-stochastic.

Then for simplicity, define $\mu = \xi_2 h_k/2$, $\nu_1 = \xi_1(2/\xi_2 - h_k)$ and $\nu_2 = \xi_1 h_k$. Thus

$$W(k) = \begin{bmatrix} (1-\mu)I & \mu I \\ \mu I - \nu_1 \bar{L}(k) & (1-\mu)I - \nu_2 \bar{L}(k) \end{bmatrix}. \tag{4.15}$$

Let λ be the eigenvalue of $W(k)$. One has

$$\begin{aligned} \det(\lambda I - W(k)) = \prod_{i=1}^{B} \big[&(\lambda^2 - 2\lambda(1-\mu) + \lambda\nu_2\gamma_i + (1-2\mu) \\ &+ (\mu\nu_2 + \mu\nu_1 - \nu_2)\gamma_i\big], \end{aligned}$$

where $\gamma_i, i = 1, 2, \ldots, B$ are the eigenvalues of $\bar{L}(k)$. Let $Q(\lambda) = \lambda^2 - 2\lambda(1-\mu) + \lambda\nu_2\gamma_i + (1-2\mu) + (\mu\nu_2 + \mu\nu_1 - \nu_2)\gamma_i$. Then $Q(1) = \mu\nu_2\gamma_i + \mu\nu_1\gamma_i$. As $0 < \mu < 1$ and $\nu_1, \nu_2 > 0$, $\lambda = 1$ implies $\gamma_i = 0$ for some i. On the other hand, if $\gamma_i = 0$: $Q(\lambda) = [\lambda - (1-2\mu)](\lambda - 1)$. Now invoking Lemmas 2.4 and 4.2, $\mathcal{G}(\bar{A}(k))$ is rooted and thus $\bar{L}(t_k)$ has one simple eigenvalue $\gamma_i = 0$. As $\mu > 0$, $1 - 2\mu \neq 1$. Hence, $\lambda = 1$ is the simple eigenvalue of $W(k)$. In view of Lemma 2.2, $\mathcal{G}(W(k))$ contains a spanning tree.

Finally, we notice that each diagonal entries of $W(k)$ is strictly positive. Hence, any vertex in $\mathcal{G}(W(k))$ has a self-arc. Recalling Lemma 2.4, the proof completes.

With the above preparations, we are now ready to present the main result:

Theorem 4.1 *Consider the digraph $\mathcal{G} = (\mathcal{V}, \mathcal{E})$. Suppose the network is under f-local attack, satisfies Assumption 4.1 and is with $(2f+1)$-robustness. If* (4.12) *holds, then the network achieves resilient consensus with rule* (4.3).

Proof *For simplicity, let us define*

$$W(k,s) \triangleq W(k)W(k-1)\dots W(s), \tag{4.16}$$

for any $k \geq s$, where $W(k,k) = W(k)$. In view of Lemma 4.3, each $W(k)$ is Sarymsakov. Since the set of Sarymsakov matrices is closed under matrix multiplication, any $W(k,s)$ belongs to the Sarymsakov class as well.

Next from (4.9), *one has*

$$y(k) = W(k,1)y(0).$$

Given Lemma 2.3, there exists some $\beta \in \mathbb{R}$ such that

$$\lim_{k\to\infty} W(k,1)y(0) = \beta \boldsymbol{1}_{2B}.$$

Therefore, for any $y_l(k), l \in \{1,2,...,2B\}$, which is the l-th entry of $y(k)$, one has

$$\lim_{k\to\infty} |y_l(k) - \beta| = 0. \tag{4.17}$$

By the definition of $y_l(k)$, one concludes that for any $i, j \in \mathcal{R}$,

$$\lim_{t\to\infty} |x_i(t) - x_j(t)| = 0. \tag{4.18}$$

Similarly, for any $i \in \mathcal{R}$,

$$\lim_{t\to\infty} |v_i(t)| = \lim_{t\to\infty} \frac{\xi_2}{2} |\tilde{v}_i(k) - \tilde{x}_i(k)| = 0. \tag{4.19}$$

The proof thus completes.

Remark 4.3 *Note in [84], where the problem is formulated in normal operation, $W(k)$ is proved to be stochastic, indecomposable and aperiodic (SIA). By further assuming that $\{W(k)\}$ belongs to a finite set, the convergence of $W(k,s)$ is guaranteed. However, since the algorithm proposed in this chapter turns to (indirectly) change the communication topology under arbitrary and misleading misbehavior, $W(k)$ is indeed drawn from an infinite set hence prevents the analysis in [84] from being directly applied. Therefore, we have drawn support from Sarymsakov matrix. Combining it with $(2f+1)$-robust network, the connectivity of $\mathcal{G}(W(k))$ is obtained. Moreover, our analysis removes the additional assumption that $h(k)$ should be chosen from a finite set.*

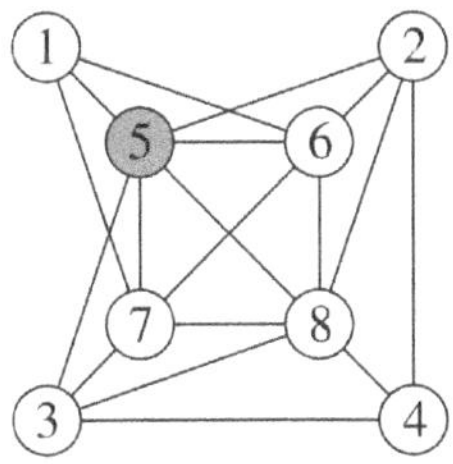

FIGURE 4.1: A three-robust communication network.

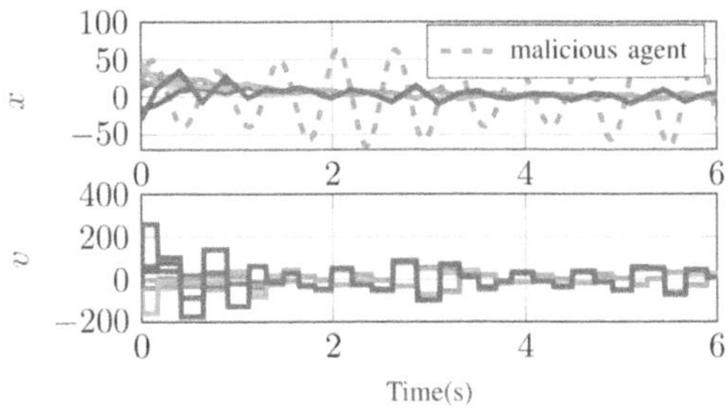

FIGURE 4.2: Trajectory of agents' states under algorithm in [84].[1]

4.6 Numerical Example

In this section, a numerical example is presented to illustrate and verify the theoretical results established above by considering the three-robust network given in [72] (see Fig. 4.1). Assume agent 5 is misbehaving. In order to deteriorate the consensus procedure, it follows a malicious way and sends false data as $x_5(t) = 50\sin(10t) + 15\cos(12t)$, which is unknown to others. On the other hand, the normal nodes always obey the proposed strategy. Firstly, Figure 4.2 presents the performance of the traditional second control consensus algorithm in [84]. The results show that the benign agents would be affected by the misbehaviors and never synchronized, necessitating resilient controllers.

To compare, we next test the algorithm proposed in this work. Let $\underline{h} = 0.15$ and $\bar{h} = 0.25$. The sampling instants are randomly chosen from $[\underline{h}, \bar{h}]$. According to (4.12), we design $\xi_2 = 5 < 2/\bar{h}$ and $\xi_1 = 0.9 < \min\{26.45, 1.5\}$. From Figure 4.3, the consensus is achieved among benign agents, regardless the faulty one.

[1]Reproduced with permission from ©2013 IEEE.

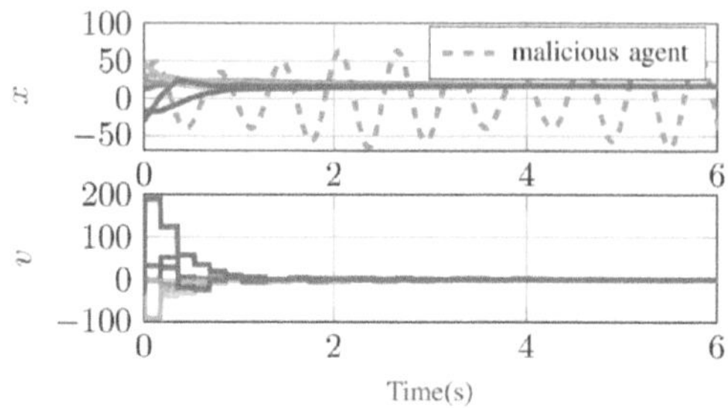

FIGURE 4.3: The Trajectory of agents' states under Algorithm 4.1.[2]

4.7 Conclusion

In this chapter, we consider the resilient consensus of second-order systems. Despite some misbehaving agents, secure protocols are desired to facilitate the agreement among normal nodes. Towards this goal, an impulsive resilient controller is presented, allowing the communication and control action occur at aperiodic sampling instants. Under certain topological conditions, such algorithm is proved to be resilient to the f-local attack. Future work will focus on the design of secure protocols via asynchronous impulsive control.

[2]Reproduced with permission from ©2013 IEEE.

5

Resilient Multi-Dimensional Consensus in Adversarial Environment

5.1 Introduction

In Chapter 4, we introduce a resilient algorithm to facilitate the agreement among benign agents in an adversarial environment. However, in some cases, the global objective is not only to facilitate consensus but also to drive the final agreement to some value related to agents' initial states. One of the most important focuses is the average consensus. Given a set of autonomous agents (such as sensors, vehicles, *etc.*), this problem seeks a distributed protocol that the agents can utilize to reach a common decision/agreement on the average of their initial values. Take an example in social networks, where each individual begins with a subjective opinion on a certain topic. Through talking with others and modifying the prejudice accordingly, the agents are supposed to agree on a *fair* view on this subject.

Although considerable attention has been paid to the development of average consensus algorithms [106, 115, 142], as reviewed in Chapter 4, these protocols are rather fragile to misbehaving agents. These agents can either prevent the network from reaching an agreement, or dictate the final consensus value on their desires [130].

Given such security concerns, this chapter considers the resilient consensus in general multi-dimensional spaces. As reported in [129], in the presence of misbehavior, no distributed rule can facilitate the exact average consensus of the benign agents' initial states. As a compromise, in this chapter, we develop a resilient algorithm such that it guarantees the agreement within the convex hull of these states. To limit the influence of network misbehavior on normal agents, a "middle points"-based protocol is proposed, where each healthy agent computes two "middle points" based on the information from its neighbors and modifies its state towards these points each time. We further show that the computation of middle points can be efficiently achieved by linear programming with a lower computational complexity. Assuming that the number of malicious agents is (locally/globally) upper bounded, sufficient

DOI: 10.1201/9781003409199-5

conditions on the network topology are presented to guarantee that the benign agents exponentially achieve the resilient consensus. Since the consensus arguably forms the foundation for distributed computing, the results in this chapter represent a first step for future works of developing resilient coordination protocols in networked systems.

5.2 Related Work

The resilient consensus has been addressed in the literature over decades. The well-known algorithms, including MSR and W-MSR, have been reviewed extensively in Section 4.2. We note, these strategies ensure the resilient consensus in uni-dimensional systems where agents' states are assumed to be scalars. The final agreement is guaranteed to be within the range limited by the minimum and maximum values of normal nodes' initial states. The implication on scalar variables, however, produces crucial limitations in various practical applications such as vehicle formation control on a 2D-plane. A naive way to generalize the results on a scalar system to a multi-dimensional system is to apply MSR or W-MSR to each entry of the state vectors. The region that the final value converges to can be immediately identified as a multi-dimensional "box." Particularly, each edge of this "box" is limited by the minimum and maximum values of benign agents' initial state in one dimension. A question thus arises naturally: is this result too conservative? Or are there any alternatives that can provide better convergence results?

To answer these questions, resilient consensus in multi-dimensional spaces has been investigated, aiming at achieving a more accurate agreement within the convex hull formed by healthy agents' initial states. In order to ensure system security, each normal node seeks a resilient convex combination each time, referring to be a point within the convex hull of its benign neighboring states. Some works achieve this through Tverberg points (see [136, 135] for examples). While the results therein are elegant, the calculation of Tverberg points is rather costly and almost impossible in many cases ([6]), and these works unfortunately do not provide an efficient way to do it. This leads to a major concern on applying Tverberg points to facilitate resilient vector consensus. This work, thereby, follows another line of research. The main contributions are summarized below:

1. Instead of Tverberg points, we propose the resilient protocol through the intersection of convex hulls. More specifically, the normal agents focus on the system dimensions alternatively and sort the received states at one dimension each time. Then each benign agent computes two "middle points" based on the sorted values, and moves its states towards an average of these points. By proving that the middle points achieve resilient convex combination, we conclude the effects of the faulty nodes on system performance are limited.

2. An explicit approach for the computation of middle points is also given, which is based on the intersection of convex hulls and can be implemented by linear programming. Note that this calculation only requires a subset of the neighboring states. Since the cardinality of the subset is fixed, the computational cost is free from the network complexity. Compared with most of the existing works, the proposed strategy is achieved with a lower computational complexity.

3. In this chapter, we are interested in the requirement on networks to achieve the resilient consensus and hereby focus on the more general incomplete graphs. Sufficient topological conditions are established to facilitate the resilient agreement. Since the idea behind canonical consensus serves as a fundamental principle in many distributed coordination settings, this method provides a powerful tool in handling misfunctioning components in multi-dimensional networked systems.

5.3 Problem Formulation

Consider a group of N agents who cooperate over the undirected graph $\mathcal{G} = \{\mathcal{V}, \mathcal{E}\}$. At any time $k \geq 0$, let $x^i(k) \in \mathbb{R}^d$ denote the current state of agent i. The agents are said to reach a (distributed) consensus if and only if there exists a constant $\bar{x}$, such that $\lim_{k\to\infty} x^i(k) = \bar{x}$ holds for every agent i. In particular, if $\bar{x} = 1/N \sum_{i=1}^{N} x^i(0)$, an average consensus is achieved.

Observe that many practical applications fit into the framework of average consensus (*e.g.*, [115, 146]). Various strategies have been developed to facilitate it in the literatures (see [103] and [106] for examples), the details of which are omitted here due to the space limitation.

5.3.1 Resilient Consensus Problem

It is worth noticing that an implicit assumption for the effectiveness of the existing approaches is that all agents are reliable throughout the execution, and cooperate to achieve the desired value. However, as the number of local agents increases, certain concerns arise that make this assumption to be violated. As discussed before, the strong dependence of distributed algorithms on the communication infrastructures creates lots of vulnerabilities for cyber attacks, where the transmitted information might be manipulated by external adversaries. Additionally, "non-participant" agent may exist, who deviates from the normal update rule and sends out self-designed information to its neighbors for its own benefits. Clearly, such misbehaviors would degrade the performance of consensus protocols: they can either prevent the benign agents from reaching a consensus, or manipulate the final agreement to be false. In fact, as shown in [130], a single "stubborn" agent can cause all agents to agree on an arbitrary value, by simply keeping this value constant.

These security concerns lead to the study of resilient consensus protocols. That is, we intend to present a secure strategy to achieve the agreement among healthy agents while raising its resilience so as to avoid being influenced by the network misbehaviors too much. By saying "resilient," we aim to achieve the following objectives, regardless of the choice of initial states and even in the adversarial environment:

1) *Agreement:* As k goes to infinity, it holds that $x^i(k) = \bar{x}$ with some $\bar{x} \in \mathbb{R}^d$, for any benign agent i;
2) *Validity:* At any time and for any benign agent, its state remains in the convex hull of all benign agents' initial values.

We elucidate these conditions as below. Firstly, the states of the benign agents should converge to the same constant value even in the presence of misbehaving ones. In addition, they are not allowed to leave the convex hull of their initial states throughout the procedure. It is observed that if a 1D problem is considered, the validity condition would be equivalent to the standard one adopted in the existing literatures [38, 64, 137, 29, 72, 32]. That is, the states of benign agents should always remain in the interval formed by the minimum and maximum of their initial values.

There has been much work proved to be effective in this simple case (*e.g.*, MSR in [38] and W-MSR in [72]). A naive way to tackle this problem in multi-dimensional spaces is by simply applying the existing scalar protocols to each component of the state vectors. Nevertheless, we should note that the region that benign agents converge to is only guaranteed as a multi-dimensional "box", each edge of which is limited by the minimum and maximum values of their initial states at one dimension. The validity condition thus fails to be ensured. To see this, we present a two-dimensional illustration in Figure. 5.1, indicating this naive way cannot guarantee the convergence to a point inside the convex hull of normal agents' initial states. Therefore, this chapter intends to address this problem and come up with an alternate method satisfying both Conditions 1) and 2).

5.3.2 Attack Model

We define $\mathcal{A}$ as the set of malicious/adversarial agents. Any agent $i \in \mathcal{A}$ could either be the adversarial one with the value being manipulated by the attacker, or the non-participant agent who does not follow the standard updating rule. On the other hand, $\mathcal{R}$ is the collection of regular/benign agents who will always follow the predefined updating strategy and compute the desired function. It is clear that $\mathcal{R} \cap \mathcal{A} = \varnothing$ and $\mathcal{R} \cup \mathcal{A} = \mathcal{V}$.

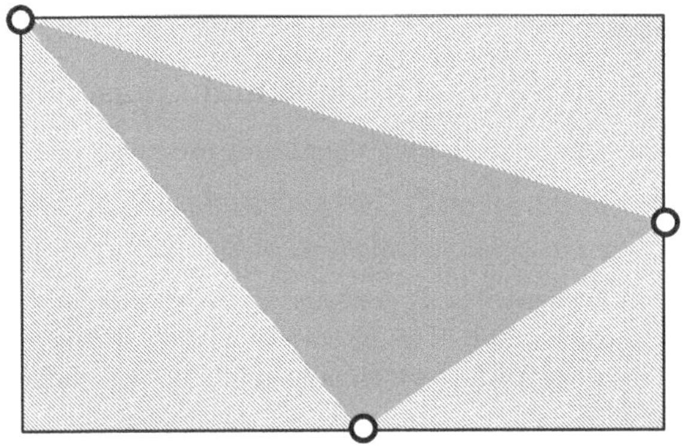

FIGURE 5.1: A 2D illustration with agents marked with circles. The location of the node indicates its initial value. With the direct application of existing algorithms to each dimension, the final agreement is ensured to be within the rectangle represented by oblique lines. However, a better solution satisfying the validity condition of converging to the solid triangle is expected.[1]

In this chapter, we characterize the network misbehaviors by the scope of threats:

1) (f-total attack model) There are at most f misbehaving agents in the network. That is, $|\mathcal{A}| \leq f$.
2) (f-local attack model) There are at most f misbehaving agents in the neighborhood of any agent. That is, $|\mathcal{A} \cap \mathcal{N}_i| \leq f$, for any agent $i \in \mathcal{V}$.

Clearly, f-total attack model is a special case of the f-local one. In the following, we shall provide different conditions to facilitate the consensus under these different models.

As before, we focus on the worst-case situation, where no restrictions are imposed on the transmitted information of agent $i \in \mathcal{A}$. That is, both adversarial and non-participant agents are allowed to send out arbitrary and different data to their neighbors. Furthermore, the faulty agents could also collude among themselves to decide on the deceptive values to be communicated.

In this chapter, we impose the following assumption on network topology:

Assumption 5.1 *For any $i \in \mathcal{V}$, it is held that $|\mathcal{N}_i| \geq (d+1)f + 1$.*

---

[1]Reproduced with permission from ©2019 Elsevier.

5.4 A Resilient Multi-Dimensional Consensus Strategy

To simplify notations, the following definitions are given beforehand:

Definition 5.1 *Consider a set $\mathcal{C} \subset \mathbb{R}^d$ with cardinality m*[2]*. For some $n \in \mathbb{Z}_{\geq 0}$ and $n \leq m$, let $\mathcal{S}(\mathcal{C}, n)$ be the set of all its subset with cardinality $m - n$.*

It is clear that the set $\mathcal{S}(\mathcal{C}, n)$ contains $\binom{m}{n}$ elements, and each of them is associated with a convex hull. The intersection of all these convex hulls plays a crucial role in our algorithm, which is formally defined below:

Definition 5.2 *Consider a set $\mathcal{C} \subset \mathbb{R}^d$ with cardinality m. For some $n \in \mathbb{Z}_{\geq 0}$ and $n \leq m$, we define $\Psi(\mathcal{C}, n)$ as*

$$\Psi(\mathcal{C}, n) \triangleq \bigcap_{S \in \mathcal{S}(\mathcal{C}, n)} \operatorname{Conv}(S). \tag{5.1}$$

In view of Definition 5.2, $\Psi(\mathcal{C}, n)$ is a subset of convex hulls formed by any $m - n$ points in $\mathcal{C}$.

5.4.1 Description of the Resilient Algorithm

In this part, we shall provide a resilient solution to the multi-dimensional consensus. Each normal agent $i \in \mathcal{R}$ starts with an initial state $x^i(0) \in \mathbb{R}^d$. At any instant $k \geq 0$, it makes updates as outlined in Algorithm 5.1.

We make some explanations on Algorithm 5.1. At each time k, the normal agents sort the p-th entry of the received values, where p varies alternatively in $\{1, 2, ..., d\}$. We refer $y^i(k)$ and $z^i(k)$ as the "middle points" in this chapter. We would prove later that these points are "safe" as they belong to the convex hull formed by benign states. Calculating them requires a subset of neighboring states and involves exactly $(d + 1)f + 1$ points in $\mathcal{X}^i(k)$. As will be shown later (in Section 5.5.4), the proposed protocol is of more lightweight than the existing ones proposed in [94, 151]. Moreover, the update law (5.2) always involves the normal agent's own state. As claimed in [72], this mechanism helps to keep more useful information at each step. Finally, as every fault-free agent is only required to access the information in its neighborhood, Algorithm 5.1 can be implemented in a distributed manner.

5.4.2 Computation of "Middle Points"

This part discusses the computation of $y^i(k)$ in Step 3, by which $z^i(k)$ can be calculated similarly. For simplicity, we omit the time index k in the sequel of this subsection.

[2]To be more precise, $\mathcal{C}$ should be defined as a multi-set since we allow duplicate elements in the set, e.g., the states of m agents shall be counted as m points even if some of them may be identical.

Algorithm 5.1 Resilient multi-dimensional consensus algorithm

1: Receive the states from all neighboring agents $j \in \mathcal{N}_i$, and collect these values in $\mathcal{X}^i(k)$.
2: Let $p = (k \bmod d) + 1$. Sort points in $\mathcal{X}^i(k)$ according to their p-th entries in an ascending order. Initialize $\mathcal{Y}^i(k)$ and $\mathcal{Z}^i(k)$ as empty.
3: Based on the sorted points, pick up the first $(d+1)f+1$ ones and collect them in $\mathcal{Y}^i(k)$. Clearly, each point in $\mathcal{Y}^i(k)$ has a smaller p-th entry than all the ones in $\mathcal{X}^i(k)\backslash\mathcal{Y}^i(k)$. Calculate *any* point $y^i(k)$, such that $y^i(k) \in \Psi(\mathcal{Y}^i(k), f)$.
4: Similarly, pick up the last $(d+1)f+1$ ones of the sorted points and collect them in $\mathcal{Z}^i(k)$. Compute any point $z^i(k) \in \Psi(\mathcal{Z}^i(k), f)$.
5: Agent i updates its local state as:

$$x^i(k+1) = \frac{x^i(k) + y^i(k) + z^i(k)}{3}. \tag{5.2}$$

6: Transmit the new state $x^i(k+1)$ to all neighbors $j \in \mathcal{N}_i$.

For simplicity, we denote $\kappa = (d+1)f+1$. Note that $\Psi(\mathcal{Y}^i, f)$ is an intersection of $r \triangleq \binom{\kappa}{f}$ convex hulls, each of which is formed by a set of $p = df+1$ points. For each of these sets, we define the matrix with the points in it as

$$Y_j = \begin{bmatrix} x^{j_1} & x^{j_2} & \cdots & x^{j_p} \end{bmatrix} \in \mathbb{R}^{d \times p}.$$

Let us denote

$$Y \triangleq \operatorname{diag}\{Y_j, j = 1, 2, \ldots, r\} \in \mathbb{R}^{dr \times pr}$$

For example, suppose $\mathcal{Y}^i = \{x^1, x^2, x^3\}$ and $f = 1$. One has

$$Y_1 = \begin{bmatrix} x^1 & x^2 \end{bmatrix}, Y_2 = \begin{bmatrix} x^1 & x^3 \end{bmatrix}, Y_3 = \begin{bmatrix} x^2 & x^3 \end{bmatrix},$$

and

$$Y = \begin{bmatrix} Y_1 & 0 & 0 \\ 0 & Y_2 & 0 \\ 0 & 0 & Y_3 \end{bmatrix}.$$

The following lemma provides an equivalent representation of $\Psi(\mathcal{Y}^i, f)$ in terms of equality and inequality constraints:

Lemma 5.1 ([140, Lemma 4]) *Let $C \in \mathbb{R}^{r \times r}$ be the circulant matrix with the first row as $\begin{bmatrix} 1 & -1 & 0 & \cdots & 0 \end{bmatrix}$. Then*

$$\Psi(\mathcal{Y}^i, f) = \left\{ \frac{1}{r} (\mathbf{1}_r' \otimes I_d) Y \beta \right\}$$

for all $\beta \in \mathbb{R}^{pr}$ such that

$$\begin{bmatrix} (C \otimes I_d) Y \\ (I_r \otimes \mathbf{1}_p') \end{bmatrix} \beta = \begin{bmatrix} \mathbf{0}_{dr} \\ \mathbf{1}_r \end{bmatrix}, \qquad \beta \geq \mathbf{0}_{pr}. \tag{5.3}$$

Note that any point in $\Psi(\mathcal{Y}^i, f)$ is acceptable. We could, thereby, choose any β^* satisfying (5.3). Then $y^i = (\mathbf{1}_r' \otimes I_n) Y \beta^* / r$. There are various manners to achieve this end. For example, Phase I method proposed in [13, Section 11.4] can be adopted by solving the following linear programming:

$$\begin{aligned} \max \quad & \alpha \\ s.t. \quad & \begin{bmatrix} (C \otimes I_d) Y \\ (I_r \otimes \mathbf{1}_p') \end{bmatrix} \beta = \begin{bmatrix} \mathbf{0}_{dr} \\ \mathbf{1}_r \end{bmatrix}, \\ & \beta_i \geq \alpha, \quad i = 1, 2, ..., pr. \end{aligned} \tag{5.4}$$

We will prove in Corollary 5.1 that $\Psi(\mathcal{Y}^i, f) \neq \varnothing$. Therefore, one can always find some β^* such that the optimal value $\alpha^* \geq 0$. The computational complexity of achieving β^* could be $\mathcal{O}\left((pr)^3\right)$ [13]. Moreover, since the cardinality of $\mathcal{Y}^i$ is fixed as $(d+1)f+1$, this computational cost is free from $|\mathcal{N}_i|$. Therefore, the algorithm will not introduce higher complexity in the network where agent i has a large number of neighbors.

Remark 5.1 *Note that, compared with the simple approach to directly use the existing scalar algorithms (like W-MSR) to each dimension of the state vectors, the proposed algorithm inevitably introduces a higher computational cost, especially when d or f is large. However, on the other hand, it would provide a better convergence result as shown in Figure 5.1. This fact indicates the trade-off between the high consensus accuracy and low computational complexity, and our work enables a fine tuning of this trade-off. To see this, one can choose to divide the d dimensions into several groups and apply our algorithm within each group. Clearly, a larger group indicates a more accurate result but a higher computational cost. In practice, it might be the case that certain components of the system variables represent more critical information. Then, one can choose to use Algorithm 5.1 on the group of these critical ones to better protect the system, while applying W-MSR to the other components to reduce computational burden.*

5.5 Algorithm Analysis

This section is devoted to the theoretical analysis of Algorithm 5.1. We shall show that the proposed algorithm is both realizable and resilient.

5.5.1 Realizability

In order to demonstrate its realizability, we shall first show the existence of $y^i(k)$ and $z^i(k)$. To this end, it is helpful to introduce Helly's Theorem as below, which is a key supporting technique of this chapter:

Helly's Theorem [27]. *Let $X_1, \ldots, X_p$ be a finite collection of convex subsets in $\mathbb{R}^d$, with $p > d$. If the intersection of every $d+1$ of these sets is nonempty, then the whole collection has a nonempty intersection. That is,*

$$\bigcap_{j=1}^{p} X_j \neq \varnothing.$$

Below is an immediate result of Helly's Theorem:

Corollary 5.1 *Let $\mathcal{C}$ be a set with cardinality m in $\mathbb{R}^d$. For any $n \in \mathbb{Z}_{\geq 0}$, if $m \geq n(d+1)+1$, then the following relation holds*

$$\Psi(\mathcal{C}, n) \neq \varnothing.$$

Proof *The result is obvious when $n = 0$. Thus we only focus on the scenario when $n \geq 1$.*

According to Definition 5.2, $\Psi(\mathcal{C}, n)$ is a intersection of $\binom{m}{n}$ convex hulls, and it is trivial to prove that $\binom{m}{n} > d$ holds. On the other hand, each of these convex hulls is created by excluding n elements of $\mathcal{C}$. Then consider any $d+1$ of them, they discard at most $n(d+1)$ points in all. Since $m \geq n(d+1)+1$, it must be the case that at least one element in $\mathcal{C}$ is retained by all of them. This indicates that any $(d+1)$ convex hulls must have a nonempty intersection. By applying Helly's Theorem, our proof is completed.

Invoking Corollary 5.1, $\mathcal{Y}^i(k), \mathcal{Z}^i(k) \neq \varnothing$. Therefore, $y^i(k)$ and $z^i(k)$ are well-defined.

5.5.2 Resiliency: Validity

To prove that Algorithm 5.1 ensures the validity condition of resilient consensus, it would be beneficial to define a set

$$\mathcal{S}^i(k) \triangleq \Psi(\mathcal{X}^i(k), f). \tag{5.5}$$

Recall that $\mathcal{X}^i(k)$ is the collection of states from $j \in \mathcal{N}_i$. In this chapter, we interpret $\mathcal{S}^i(k)$ as a "safe kernel" as illustrated in Figure 5.2.

The below results would be helpful:

Lemma 5.2 *Consider two collections of sets $\{A_i\}_{i \in I}$ and $\{B_j\}_{j \in \mathcal{J}}$. If for any $j \in \mathcal{J}$, there exists an $i^* \in \mathcal{I}$ such that $A_{i^*} \subset B_j$, then*

$$\bigcap_{i \in \mathcal{I}} A_i \subset \bigcap_{j \in \mathcal{J}} B_j.$$

Proof *Denote a subset of $\{A_i\}_{i \in I}$ as $\{A_{i^*}\}_{i^* \in I}$, such that $\{A_{i^*}\}_{i^* \in I}$ contains all A_{i^*} which has a superset in $\{B_j\}_{j \in \mathcal{J}}$. The proof is then completed by noticing that*

$$\bigcap_{i \in \mathcal{I}} A_i \subset \bigcap_{i^* \in \mathcal{I}} A_{i^*} \subset \bigcap_{j \in \mathcal{J}} B_j.$$

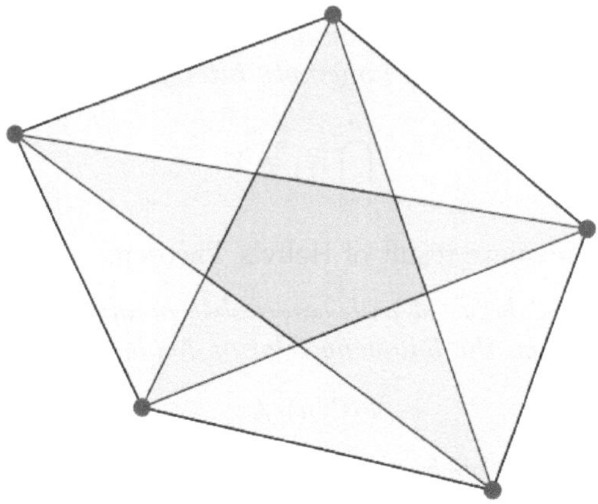

FIGURE 5.2: A 2D illustration of "safe kernel." Suppose agent $i \in \mathcal{R}$ has five neighbors and each of their states is represented by the location of a circle. Let $f = 1$. The green region denotes the safe kernel $\mathcal{S}^i(k) = \Psi(\mathcal{X}^i(k), 1)$.[3]

Lemma 5.3 *Consider any set $\mathcal{C}_1$ with cardinality m_1 and $\mathcal{C}_2$ with cardinality m_2. If $\mathcal{C}_1 \subset \mathcal{C}_2$, then for any $n \leq m_1$, the following statement holds:*

$$\Psi(\mathcal{C}_1, n) \subset \Psi(\mathcal{C}_2, n).$$

Proof *We shall prove that every set S_2 in $\mathcal{S}(\mathcal{C}_2, n)$ is a superset for some set S_1 in $\mathcal{S}(\mathcal{C}_1, n)$. To see this, notice that*

$$\begin{aligned} S_2 = \mathcal{C}_2 \backslash S_2^c &\supset (\mathcal{C}_2 \backslash S_2^c) \cap \mathcal{C}_1 \\ &= \mathcal{C}_1 \backslash (S_2^c \cap \mathcal{C}_1), \end{aligned}$$

where $S_2^c = \mathcal{C}_2 \backslash S_2$ is a set with cardinality n. Notice that $S_2^c \cap \mathcal{C}_1$ has cardinality no greater than n, which means that $\mathcal{C}_1 \backslash (S_2^c \cap \mathcal{C}_1)$ is a superset of some set in $\mathcal{S}(\mathcal{C}_1, n)$. The proof is thus finished by invoking Lemma 5.2.

From Algorithm 5.1, it is clear that $\mathcal{Y}^i(k), \mathcal{Z}^i(k) \subset \mathcal{X}^i(k)$. In view of Lemma 5.3, one has $y^i(k), z^i(k) \in \mathcal{S}^i(k)$, namely the middle points are always within the safe kernel.

Now consider the validity condition of resilient consensus. For simplicity, we denote the convex hull formed by the benign agents' states at time k as $\Omega(k)$. The following proposition presents the non-expansion property of $\Omega(k)$:

Proposition 5.1 (Validity) *Consider the network $\mathcal{G} = (\mathcal{V}, \mathcal{E})$. Suppose the misbehaving agents follow either f-local or f-total attack model. With Algorithm 5.1, the following relation holds at any $k \geq 0$:*

$$\Omega(k+1) \subset \Omega(k). \tag{5.6}$$

[3]Reproduced with permission from ©2019 Elsevier.

Proof *Consider the scenario under either f-local or f-total attacks. For a benign agent i, there exist no less than $|\mathcal{X}^i(k)| - f$ benign ones in its neighborhood. By definitions, one obtains that $\mathcal{S}^i(k)$ is included in the convex hull formed by any $|\mathcal{X}^i(k)| - f$ neighboring values. Hence, it is trivial to derive that $\mathcal{S}^i(k)$ is a subset of the convex hull of the benign neighbors' states, that is, $\mathcal{S}^i(k) \subset \Omega(k)$. Recall (5.2), one directly has that $x^i(k+1) \in \Omega(k)$ as it is a convex combination of some points in $\Omega(k)$.*

Because the above relation holds for any normal node, one has $\Omega(k+1) \subset \Omega(k)$.

Hence, the safe kernel is "safe" in the sense that it is contained in the convex hull formed by benign states. With two middle points in it, Algorithm 5.1 guarantees the validity condition of resilient consensus. That is, the healthy agents would never move out of the convex hull formed by their initial values, namely $\Omega(0)$, despite the influence of the misbehaving ones.

5.5.3 Resiliency: Agreement

We first introduce the below lemma:

Lemma 5.4 *Let $\mathcal{C}$ be a set with $|\mathcal{C}| = (d+1)n+1$. The following relations hold for any linear function $l(x)$:*

1) *If there exists at least $dn+1$ points $\bar{x}$ in $\mathcal{C}$, such that $l(\bar{x}) \leq m$, then for any point $y \in \Psi(\mathcal{C}, n)$, $l(y) \leq m$ holds;*
2) *If there exists at least $dn+1$ points $\bar{x}$ in $\mathcal{C}$, such that $l(\bar{x}) \geq M$, then for any point $z \in \Psi(\mathcal{C}, n)$, $l(z) \geq M$ holds.*

Proof *By Corollary 5.1, $\Psi(\mathcal{C}, n) \neq \varnothing$. We then show the rationale of the statements as follows:*

1) *There exist $dn+1$ points $\bar{x}$ in $\mathcal{C}$ such that $l(\bar{x}) \leq m$. Consider the convex hull formed by these $\bar{x}$'s. Clearly, any element x in this convex hull also has $l(x) \leq m$. We could infer from Definition 5.2 that $\Psi(\mathcal{C}, n)$ is a subset of this convex hull, and thus the first statement holds.*
2) *The second statement is proved in a similar manner as above.*

For simplicity, let $l_p(x) = e_p^T x$, a linear function that returns the p-th entry of $x \in \mathbb{R}^d$. Then among all $l_p(x)$'s, where $x \in \mathcal{X}^i(k)$, we respectively denote $\underline{m}_p(k)$ and $\overline{M}_p(k)$ as the $(df+1)$-th smallest and largest values. From Lemma 5.4, the next result gives the bounds on middle points:

Corollary 5.2 *Consider the network $\mathcal{G} = (\mathcal{V}, \mathcal{E})$. Suppose normal agent $i \in \mathcal{R}$ updates by Algorithm 5.1, then it holds that:*

$$\begin{aligned} y_p^i(k) &\leq \underline{m}_p(k), \\ z_p^i(k) &\geq \overline{M}_p(k). \end{aligned} \tag{5.7}$$

Now it is ready to provide the main results. In what follows, we shall present sufficient conditions on network topology, under which the agreement condition of resilient consensus will be guaranteed:

Proposition 5.2 (Agreement) *Consider the network $\mathcal{G} = (\mathcal{V}, \mathcal{E})$. Suppose the misbehaving agents follow an f-local attack model. If the network is with $((d+1)f+1)$-robustness, then with Algorithm 5.1, all the benign agents are guaranteed to achieve consensus exponentially, regardless of the actions of misbehaving agents.*

Proof *To proceed, at any $p \in \{1, 2, ..., d\}$, let us respectively denote $m_p(k)$ and $M_p(k)$ as the minimum and maximum value among the p-th components of normal agents' states at time k. That is,*

$$\begin{aligned} m_p(k) &\triangleq \min_{i \in \mathcal{R}} x_p^i(k), \\ M_p(k) &\triangleq \max_{i \in \mathcal{R}} x_p^i(k). \end{aligned} \tag{5.8}$$

As a direct result of Proposition 5.1, $m_p(k+1) \geq m_p(k)$ and $M_p(k+1) \leq M_p(k)$ for any p and k.

To establish the achievement of consensus, it is equivalent to prove that the benign agents reach an agreement at any dimension. Due to the symmetry between different dimensions, without loss of generality, we would only focus on the first component of local states. The temporal difference is thereby defined as $\Delta_1(k) = M_1(k) - m_1(k)$. We attempt to show that $\Delta_1(k)$ asymptotically approaches 0.

For notation convenience, the following definitions are further imposed for any $\bar{k} \geq k$, and any $\epsilon \in \mathbb{R}$:

$$\begin{aligned} \mathcal{V}^M(k, \bar{k}, \epsilon) &\triangleq \{i \in \mathcal{V} : x_1^i(\bar{k}) > M_1(k) - \epsilon\}, \\ \mathcal{V}^m(k, \bar{k}, \epsilon) &\triangleq \{i \in \mathcal{V} : x_1^i(\bar{k}) < m_1(k) + \epsilon\}. \end{aligned} \tag{5.9}$$

Note that the subscript is dropped in the above notations for the sake of brevity. Clearly, $\mathcal{V}_M(k, \bar{k}, \epsilon)$ [resp. $\mathcal{V}_m(k, \bar{k}, \epsilon)$] includes all agents whose state's first component is greater [resp. less] than $M(k) - \epsilon$ [resp. $m(k) + \epsilon$] at time $\bar{k}$. We then define

$$\begin{aligned} \mathcal{R}^M(k, \bar{k}, \epsilon) &\triangleq \mathcal{V}^M(k, \bar{k}, \epsilon) \cap \mathcal{R}, \\ \mathcal{R}^m(k, \bar{k}, \epsilon) &\triangleq \mathcal{V}^m(k, \bar{k}, \epsilon) \cap \mathcal{R}, \end{aligned} \tag{5.10}$$

which contains only benign agents in $\mathcal{V}^M(k, \bar{k}, \epsilon)$ and $\mathcal{V}^m(k, \bar{k}, \epsilon)$, respectively.

Suppose that $M_1(k) \neq m_1(k)$, i.e., $\Delta_1(k) > 0$ at some time step k such that $p = (k \bmod d) + 1 = 1$. Define $\epsilon_0 = \Delta_1(k)/2$. It is easy to know that $\mathcal{R}^M(k, k, \epsilon_0)$ and $\mathcal{R}^m(k, k, \epsilon_0)$ are disjoint. Furthermore, since each of these sets contains a benign agent with the first component being $M_1(k)$ or $m_1(k)$, both of them are nonempty. As the network is $((d+1)f+1)$-robust, there *exists one benign node in either* $\mathcal{R}^M(k,k,\epsilon_0)$ *or* $\mathcal{R}^m(k,k,\epsilon_0)$ *that has at least* $(d+1)f+1$ *neighboring agents outside its set.*

Without loss of generality, let $i \in \mathcal{R}^M(k,k,\epsilon_0)$ *be such an agent who has no less than* $(d+1)f+1$ *neighbors in* $\mathcal{V}\backslash\mathcal{R}^M(k,k,\epsilon_0)$*. Moreover, under the* f*-local attack model, no less than* $df+1$ *points in agent* i*'s neighborhood have their first components upper bounded by* $M_1(k)-\epsilon_0$*. Therefore, one has that* $\underline{m}_1(k) \le M_1(k)-\epsilon_0$*.*

Invoking Corollary 5.2, $y_1^i(k) \le \underline{m}_1(k) \le M_1(k)-\epsilon_0$*. We thus obtain that*

$$\begin{aligned} x_1^i(k+1) &= \frac{1}{3}(x_1^i(k)+y_1^i(k)+z_1^i(k)) \\ &\le \frac{2}{3}M_1(k)+\frac{1}{3}(M_1(k)-\epsilon_0) \\ &= M_1(k)-\frac{1}{3}\epsilon_0. \end{aligned} \tag{5.11}$$

It is pointed out that this upper bound also applies to any benign agent in $\mathcal{V}\backslash\mathcal{V}^M(k,k,\epsilon_0)$*, as it will apply its own value for updates.*

Similarly, if the benign agent $j \in \mathcal{R}^m(k,k,\epsilon_0)$ *has at least* $(d+1)f+1$ *neighbors outside its set, we know that* $z_1^i(k) \ge \overline{M}^l \ge m_1(k)+\epsilon_0$ *and shall have an analogous result that* $x_1^j(k+1) \ge m_1(k)+\epsilon_0/3$*, which again, is the lower bound for every benign agent in* $\mathcal{V}\backslash\mathcal{V}^m(k,k,\epsilon_0)$*.*

Define $\epsilon_1 = \epsilon_0/3$*. From former discussions, one knows that at least one benign agent in* $\mathcal{R}^M(k,k,\epsilon_0)$ *has its first component decreased to below* $M_1(k)-\epsilon_1$*, or one benign agent in* $\mathcal{R}^m(k,k,\epsilon_0)$ *has its first component increased to above* $m_1(k)+\epsilon_1$*, or both. As a result, it must be either* $\mathcal{R}^M(k,k+1,\epsilon_1) \subsetneq \mathcal{R}^M(k,k,\epsilon_0)$*, or* $\mathcal{R}^m(k,k+1,\epsilon_1) \subsetneq \mathcal{R}^m(k,k,\epsilon_0)$*, or both.*

Then consider the update at $k+2$*. For any normal agent in* $\mathcal{V}\backslash\mathcal{V}^M(k,k+1,\epsilon_1)$*, it is trivial to see that*

$$\begin{aligned} x_1^i(k+2) &= \frac{1}{3}(x_1^i(k+1)+y_1^i(k+1)+z_1^i(k+1)) \\ &\le \frac{1}{3}(M_1(k)-\epsilon_1)+\frac{2}{3}M_1(k+1) \\ &\le \frac{1}{3}(M_1(k)-\epsilon_1)+\frac{2}{3}M_1(k) \\ &= M_1(k)-\epsilon_2, \end{aligned} \tag{5.12}$$

with $\epsilon_2 = \epsilon_1/3$*. Thereby,* $\mathcal{R}^M(k,k+2,\epsilon_2) \subset \mathcal{R}^M(k,k+1,\epsilon_1)$*. Hence, for each* $1 \le t \le d$*, we can recursively define* $\epsilon_t = \epsilon_0/3^t$ *and obtain that* $\mathcal{R}^M(k,k+d,\epsilon_d) \subset \mathcal{R}^M(k,k+1,\epsilon_1) \subsetneq \mathcal{R}^M(k,k,\epsilon_0)$*. Similarly,* $\mathcal{R}^m(k,k+d,\epsilon_d) \subsetneq \mathcal{R}^m(k,k,\epsilon_0)$*.*

Note that $\epsilon_d < \epsilon_0$*, and therefore the sets* $\mathcal{R}^M(k,k+d,\epsilon_d)$ *and* $\mathcal{R}^m(k,k+d,\epsilon_d)$ *are still disjoint. If both sets are nonempty, as above, one can conclude that at least one of the following statements is true:* $\mathcal{R}^M(k,k+2d,\epsilon_{2d}) \subsetneq \mathcal{R}^M(k,k+d,\epsilon_d)$*, or* $\mathcal{R}^m(k,k+2d,\epsilon_{2d}) \subsetneq \mathcal{R}^m(k,k+d,\epsilon_d)$*.*

Hence, for any $\kappa \geq 1$*, as long as both* $\mathcal{R}^M(k, k+\kappa d, \epsilon_{\kappa d})$ *and* $\mathcal{R}^m(k, k+\kappa d, \epsilon_{\kappa d})$ *are nonempty, we can repeat the above analysis and conclude that at least one of these two sets will shrink at the next time step. Since* $|\mathcal{R}^M(k, k, \epsilon_0)| + |\mathcal{R}^m(k, k, \epsilon_0)| \leq |\mathcal{R}|$*, there must be the case that either* $\mathcal{R}^M(k, k+d|\mathcal{R}|, \epsilon_{d|\mathcal{R}|}) = \varnothing$*, or* $\mathcal{R}^m(k, k+d|\mathcal{R}|, \epsilon_{d|\mathcal{R}|}) = \varnothing$*, or both. We assume the former statement holds. From (5.9), at time step* $k + d|\mathcal{R}|$*, all the fault-free agents have their first elements being at most* $M_1(k) - \epsilon_{d|\mathcal{R}|}$*, i.e.,* $M_1(k + d|\mathcal{R}|) \leq M_1(k) - \epsilon_{d|\mathcal{R}|}$*. On the other hand, from Proposition 5.1, we have* $m_1(k + d|\mathcal{R}|) \geq m_1(k)$*. As a result,*

$$\Delta_1(k + d|\mathcal{R}|) \leq \Big(1 - \frac{1}{2 \cdot 3^{d|\mathcal{R}|}}\Big)\Delta_1(k). \tag{5.13}$$

Therefore we conclude that $\Delta_1(k)$ *vanishes exponentially.*

The next result elaborates a different condition for the proposed algorithm to succeed in f-total threats:

Proposition 5.3 (Agreement) *Consider the network* $\mathcal{G} = (\mathcal{V}, \mathcal{E})$*. Suppose the misbehaving agents follow an* f*-total attack model. If the network is with* $(df+1, f+1)$*-robustness, then with Algorithm 5.1, all the benign agents are guaranteed to achieve consensus exponentially, regardless of the actions of misbehaving agents.*

Proof *Proposition 5.3 is proved in a similar manner to that of Proposition 5.2. The essential point is that if* $\mathcal{V}^M(k, k+\kappa d, \epsilon_{\kappa d})$ *and* $\mathcal{V}^m(k, k+\kappa d, \epsilon_{\kappa d})$ *are nonempty and disjoint, and if both of these sets contain some benign agents, then under* $(df+1, f+1)$*-robust graph, there exists at least one benign agent in either* $\mathcal{V}^M(k, k+\kappa d, \epsilon_{\kappa d})$ *or* $\mathcal{V}^m(k, k+\kappa d, \epsilon_{\kappa d})$ *that has no less than* $df+1$ *neighboring agents outside its set. Suppose the benign agent* $i \in \mathcal{V}^M(k, k+\kappa d, \epsilon_{\kappa d})$ *is such a node. Following a similar proof procedure of Proposition 5.2, we know that agent* i *will always apply a state that has its first entry being no more than* $M_1(k) - \epsilon_{\kappa d}$ *for updating under Algorithm 5.1. The result can be finally concluded by applying the proof techniques as before.*

Remark 5.2 *By definitions, we note that a* $((d+1)f+1)$*-robust graph is* $(df+1, f+1)$*-robust as well, but not vice versa. That is to say, the network which is able to tolerate* f*-local attacks could also survive the* f*-total ones, while the converse is not true. This observation is consistent with the fact that the* f*-globally bounded threats are special versions of locally bounded ones.*

Given the above results, one thus immediately concludes that the proposed algorithm facilitates the resilient consensus, as stated in the following:

Theorem 5.1 *Consider the network* $\mathcal{G} = (\mathcal{V}, \mathcal{E})$*. Suppose the network satisfies one of the following conditions:*
1) under f*-local attack model, and is* $((d+1)f+1)$*-robust,*
2) under f*-total attack model, and is* $(df+1, f+1)$*-robust.*

With Algorithm 5.1, benign agents exponentially achieve the resilient consensus, regardless of the actions of misbehaving ones. That is, as $k \to \infty$*,*

$$x^i(k) = x^j(k) = \hat{x} \quad \text{for any } i, j \in \mathcal{R}, \tag{5.14}$$

where $\hat{x} \in \Omega(0)$*.*

Proof *The theorem is immediately achieved as both the validity and agreement conditions have been established in Propositions 5.1–5.3.*

Theorem 5.1 indicates that under certain topological conditions, the safe kernel approach guarantees all benign agents reach an agreement on a weighted average of their initial states. It protects the local states of benign agents from being driven to arbitrary values and thus could withstand the compromise of partial agents while providing a desired level of security. Furthermore, since its convergence does not depend on the actions of misbehaving agents, it works effectively even in the worst-case scenario, where the faulty agents could have full knowledge of graph topology, updating rules, etc., and could also be Byzantine agents that are able to send different information to different neighbors. Finally, let us consider the scenario when there is no faulty agents at all, *i.e.*, $|\mathcal{F}| = 0$. Theorem 5.1 shows that the proposed strategy only guarantees a 'decent' solution (i.e., within $\Omega(0)$) instead of the exact average value. This implies that in order to increase the system's resilience, we sacrifice its performance during normal operations. Hence, there exists a trade-off between security and optimality.

5.5.4 Remarks on the Safe Kernel

We finally make some remarks on the safe kernel. Consider the consensus procedure. At every step, the normal agent i obtains the states in its neighborhood, whereas up to f of them might be faulty. To ensure its state updated in a safe manner, agent i hopes to use only good inputs. Yet as it has no knowledge on the identities of these values, it intends to achieve a *resilient convex combination*, which lies in the convex hull of any $|\mathcal{N}_i| - f$ neighboring states.

Different from [136, 135], where Tverberg points are applied, this chapter, inspired by Helly's Theorem, achieves the resilient convex combination through the intersection of convex hulls. Let us recall the safe kernel $\mathcal{S}^i(k)$ illustrated in Figure 5.2. From Definitions 5.1 and 5.2, it intuitively ignores the effects from combination of any f values. At any time, the healthy agent modifies its state towards this kernel. The impacts of malicious agents on the benign ones are thus limited, with formal proof given in Proposition 5.1.

Note that, in [94, 151], the safe kernel $\mathcal{S}^i(k)$ is required to be exactly calculated. Although the results therein are elegant, a major concern of applying them is the high-computational complexity. More specifically, $\mathcal{S}^i(k)$ is the intersection of $\binom{|\mathcal{N}_i|}{f}$ convex hulls, and existing approaches for the computation of this intersection are usually #p-hard ([110]). Moreover, since this computation depends on every state in $\mathcal{X}^i(k)$, these algorithms may fail in a large neighborhood where node i has a large number of neighbors. As a contrast, although the proof of this work is established on $\mathcal{S}^i(k)$, we indeed do not need to compute the whole kernel, but two "middle points" insides. As shown in Section 5.4.2, this step can be achieved by a linear programming, the computational complexity of which will not increase within a larger neighborhood.

5.6 Discussions on the Network Failing to Meet Suffcient Conditions

Observe that in Assumption 5.1, we assume the network is with large connectivity so that every agent has a sufficient number of neighbors. Furthermore, the resiliency analysis shows that the network robustness required to achieve the consensus increases linearly with the dimension of the agents' state. Thus, a natural question arises: what if the network is not "connected" or "robust" enough to meet these assumptions? This section is devoted to investigating this issue. To highlight the results, we focus on the standard consensus settings.

A naive way to handle this problem works as follows. Suppose the given network can tolerate f (locally/globally) faulty agents only when the system dimensionality is no more than $d'(< d)$. Then one could group every d' of the d components together (if d is not divisible by d', then there is a single group whose cardinality is strictly less than d'.). At any time step, Algorithm 5.1 is applied within every group by every benign agent. The updated results will then be rejoined in order as a d-dimensional vector to be broadcast to its neighbors. It is worth pointing out that this approach fails to guarantee the validity condition but instead only restricts the achieved agreement within the convex hull on every d' (or less) dimensions in a group. Particularly, if $d' = 1$, its performance will degrade to that of directly applying W-MSR.

The above attempt is based on the existing network and is rather straightforward. Another possible solution is by adding some well-designed "trusted" agents in the network. The trusted nodes are the ones who are well protected and cannot be compromised by any attacker. If we denote the set of such agents as $\mathcal{T}$. Clearly $\mathcal{T} \subset \mathcal{R}$.

Then let us consider the scenario where the network holds a subset of "trusted" agents. As before, each normal agent needs to create a safe kernel based on its neighboring states. However, this procedure could be simplified with its trusted neighbors, as listed in Algorithm 5.2.

As indicated in this algorithm, if the trusted neighbors exist, the middle points are established only by their states. Hence we can relax Assumption 5.1 to be:

Assumption 5.2 *For any $i \in \mathcal{V}$, it is held that either $\mathcal{T} \cap \mathcal{N}_i \neq \varnothing$ or $|\mathcal{N}_i| \geq (d+1)f+1$.*

Algorithm 5.2 Resilient multi-dimensional consensus algorithm in the presence of trusted agents

---

1: Receive the states from all neighboring agents $j \in \mathcal{N}_i$, and collect these values in $\mathcal{X}^i(k)$.
2: Let $p = (k \bmod d) + 1$. Sort points in $\mathcal{X}^i(k)$ according to their p-th entries in an ascending order.
3: If $\mathcal{T} \cap \mathcal{N}_i \neq \varnothing$, that is, agent i has at least one trusted neighbor, let $y^i(k)$ and $z^i(k)$ respectively be the states of trusted neighbors who have the smallest and largest p-th entries. Otherwise, calculate $y^i(k)$ and $z^i(k)$ as in Steps 3–4 of Algorithm 5.1.
4: Agent i updates its local state as:

$$x^i(k+1) = \frac{x^i(k) + y^i(k) + z^i(k)}{3}. \tag{5.15}$$

5: Transmit the new state $x^i(k+1)$ to all neighbors $j \in \mathcal{N}_i$.

With the above notions, the resilience result can be established as below:

Theorem 5.2 *Consider the network $\mathcal{G}(\mathcal{V}, \mathcal{E})$. Suppose the network satisfies one of the following conditions:*
1) under f-local attack, and is $((d+1)f+1)$-robust with $\mathcal{T}$,
2) under f-total attack, and is $(df+1, f+1)$-robust with $\mathcal{T}$.
With Algorithm 5.2, benign agents exponentially achieve the resilient consensus, regardless of the actions of misbehaving ones.

Proof *This proof is established in an analogous manner as before. The major difference is as follows:*

Consider f-local attack model. Suppose both $\mathcal{R}^M(k, k+t, \epsilon_t)$ and $\mathcal{R}^m(k, k+t, \epsilon_t)$ are nonempty. There exists $i \in \mathcal{R}^M(k, k+t, \epsilon_t) \cup \mathcal{R}^m(k, k+t, \epsilon_t)$ such that it either has no less than $(d+1)f+1$ neighbors outside its respective set, or has a trusted neighbor outsides. Hence it always applies a state, namely $y^i(k)$, which has its first entry being no more than $M_1(k) - \epsilon_t$ for updating. One could develop similar results in an f-total attack scenario. The whole proof is however omitted due to the space limitation.

Remark 5.3 *In special scenarios where $\mathcal{T}$ forms a connected dominating set,*[4] *the network is with any robustness with $\mathcal{T}$. According to Theorem 5.2, it can tolerate any number of misbehaving nodes and achieve the resilient consensus in multi-dimensional spaces. We will further study this case in Section 5.7.*

---

[4]A set $\mathcal{T} \subset \mathcal{V}$ is a connected dominating set if 1) For any $i \in \mathcal{V}$, there exists some $j \in \mathcal{T}$ such that $j \in \mathcal{N}_i$; and 2) $\mathcal{T}$ induce a connected graph [4].

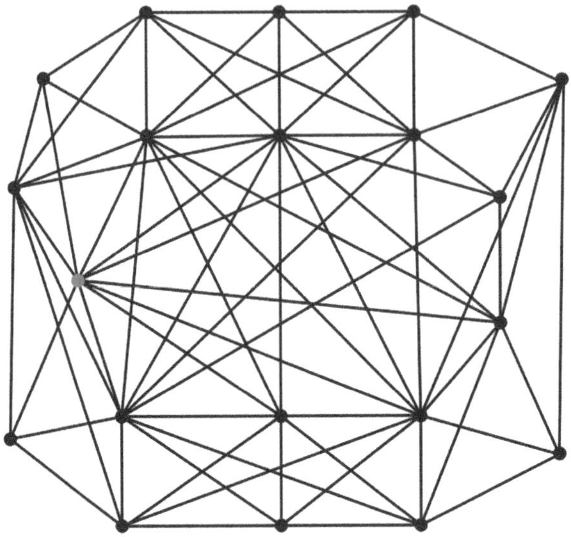

FIGURE 5.3: A $(3, 2)$-robust communication network.[5]

5.7 Numerical Example

In this section, we provide a numerical example to verify the theoretical results established in the previous sections.

The $(3, 2)$-robust graph in Figure 6.1 is chosen as the communication network,[6] where the node set is $\mathcal{V} = \{1, 2, \ldots, 20\}$. Proposition 5.3 indicates that the network can tolerate a single misbehaving node in a two-dimensional system. This agent intends to prevent others from reaching a correct consensus by violating the rule in Algorithm 5.1 and setting its states as $x_1^2(k) = 4.5 * \sin(k/5)$ and $x_2^2(k) = k/25 + 1$ at any time k. On the other hand, the benign agents are initialized randomly within $[-3, 3] \times [0, 5]$ and always follow (5.2) for updates.

The performance of Algorithm 5.1 is tested in Figures 5.4 and 5.5. The result shows that the states of benign agents are guaranteed to be within the convex hull of their initial states at any time and they finally achieve a common value, which validates the established results. Since the malicious

[5]Reproduced with permission from ©2019 Elsevier.

[6]This network is established based on the $(3, 2)$-robust network given in [73] and Theorem 1 therein, which shows how to construct an (r, s)-robust digraph with an existing (r, s)-robust one.

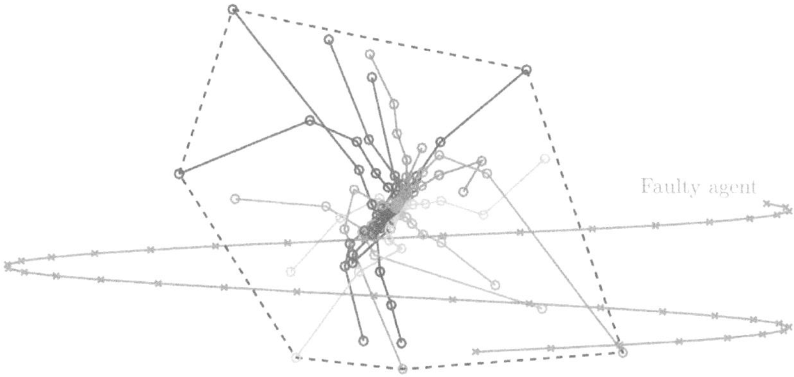

FIGURE 5.4: The trajectory of local states under Algorithm 5.1, where the area surrounded by the dashed lines is the convex hull of the initial states of benign agents.[7]

agent is unable to affect the final agreement too much, our protocol helps to improve the system security.

5.8 Conclusion

Due to their wide applications, the distributed coordination in networked systems has attracted much research interest. In this work, we are interested in the achievement of consensus under malicious agents. A resilient "middle points" based algorithm is proposed in this work, which is also applicable in the high-dimensional spaces. By solving the middle points through a linear programming, the proposed strategy introduces a lower computational complexity than most of the existing works. Under certain network topology, it guarantees the benign agents exponentially reach an agreement within the convex hull of their initial states. Finally, as the consensus arguably forms the foundation for distributed computing, the results in this chapter lay a solid foundation for future works to develop resilient coordination protocols in other consensus-based problems.

[7]Reproduced with permission from ©2019 Elsevier.

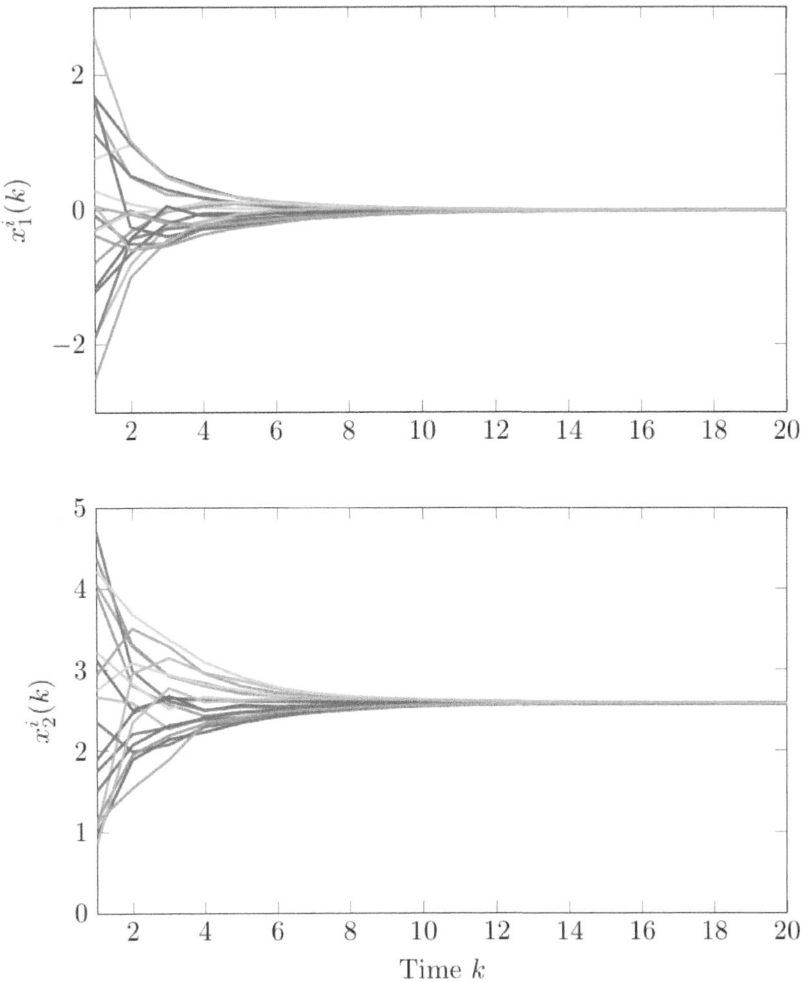

FIGURE 5.5: The states of normal agents $x^i(k), \forall i \in \mathcal{R}$.[8]

[8]Reproduced with permission from ©2019 Elsevier.

6

Resilient Containment Control in Adversarial Environment

6.1 Introduction

This chapter investigates another important problem in networked systems, namely containment control. The existing consensus problems often focus on the leaderless coordination of agents. However, in practice, there might exist one or multiple leaders for these agents. Distributed coordination with a single leader has been widely discussed in the leader-following cases [50, 104, 162]. On the other hand, containment control problems are proposed to address the scenario where a group of followers is guided by multiple leaders. Instead of facilitating the consensus, such problems seek for appropriate distributed algorithms that the followers could utilize to move into the convex hull spanned by the leaders.

In this chapter, we consider the problem of containment control in an adversarial environment, where some of the agents (who can be either leaders or followers) are misbehaving. Despite the influence of network misbehaviors, the normal followers aim to move towards the convex hull formed by the normal leaders. To this end, resilient containment control is investigated in this work. We propose secure protocols for both first-order and second-order systems, where each normal follower ignores the most extreme values in its neighborhood and modifies its state based on the remaining ones. Sufficient conditions related to the network topology and the maximum number of tolerable faulty nodes are finally derived to guarantee the achievement of resilient containment. Numerical examples are also provided in the end to verify our theoretical results.

6.2 Related Work

Study of containment control is motivated by the natural phenomena and has potential and important applications in practice [16, 77, 47]. For example,

DOI: 10.1201/9781003409199-6

the male silkworm moths will end up within the convex hull spanned by the female silkworm moths by detecting pheromone released by females. Another example involves a group of vehicles where only some of them are equipped with necessary sensors to detect obstacles. The vehicles with these sensors can be regarded as "leaders" while the others are "followers." The leaders first form a safety area based on their knowledge of obstacles. Then, through containment control, the followers can stay within the convex hull formed by the leaders and thus remain safe as well.

Since the early work [58], where a Stop-Go policy is proposed to realize the containment control under a fixed topology, much research effort has been devoted to this field (see [95, 92, 81, 77, 37, 82]). Given different system dynamics, the containment control of linear and Euler-Lagrangian systems are respectively considered in [77, 92]. Apart from stationary leaders, the containment control in the presence of multiple dynamics leaders is considered in [74] as well. In addition, Meng et al. [95] propose a finite-time realization to achieve a faster convergence. It is worth noticing that the existing works are proposed in benign environment where both computing agents and communication channels are trustworthy enough. Nevertheless, the increasing vulnerabilities of networks to both internal and external misbehaviors call for the resilient protocols working efficiently in adversarial environment.

As reviewed in Chapters 4 and 5, resilient control of networked systems has been studied over decades. However, despite the elegant results achieved regarding resilient consensus and optimization (see [69, 38, 72, 130, 133] for examples), few research efforts have been devoted to the security of containment control. Hence we address this issue and seek a secure protocol to guarantee the normal followers move to the convex hull formed by the benign leaders even in adversarial environment. As a start work, this chapter focuses on simple scalar systems. The main contributions are two-fold:

1. Resilient containment control is investigated for both first-order and second-order systems. To enhance its security, we adopt the similar idea to most resilient strategies. That is, at each time, a normal follower removes the most extreme values in its neighborhood and creates a *safe region*. Only the values in this region would be applied in the subsequent updating stage. We provide sufficient conditions on network topology such that the proposed strategy always results in a "virtual" network with a united spanning tree,[1] regardless of the misbehaviors.

2. Through the convexity analysis and using a Lyapunov function, we prove that the largest distance from the normal followers to the convex hull spanned by the normal leaders will converge to 0. Therefore, the resilience of the proposed strategies is guaranteed.

[1]The digraph is said to have a united spanning tree if for each of the follower nodes, there exists at least one leader node that has a directed path to it [77].

6.3 Problem Formulation

Let us consider a group of n agents. These agents cooperate over the network modeled by a digraph $\mathcal{G} = \{\mathcal{V}, \mathcal{E}\}$. We define the agents which do not have any in-neighbors as leaders, while the others as followers. The sets of leaders and followers are respectively denoted by $\mathcal{L}$ and $\mathcal{F}$. The target of containment control is to drive the followers into the convex hull spanned by the multiple leaders.

6.3.1 Attack Model

This chapter considers the containment control in an adversarial environment where some of the agents might be faulty or misbehaving. Such misbehaving ones can deteriorate the system performance by misleading the normal followers to leave the convex hull of normal leaders or destroying the convex formation of leaders.

With the same notations as in Chapter 5, let us denote the set of faulty or misbehaving agents as $\mathcal{A}$. Any agent $i \in \mathcal{A}$ could either be the one whose value is manipulated by the attacker, or the one who fails to follow the standard updating rule. On the other hand, the normal or benign agents will always adopt the prescribed control strategy, and are collected in the set $\mathcal{R}$. Both the leaders and followers can be normal or adversarial. For notation convenience, we denote the normal leaders and followers by $\mathcal{L}_\mathcal{R}$ and $\mathcal{F}_\mathcal{R}$ respectively.

This chapter still assumes the f-local attack model given in Definition 4.1. We aim to develop control algorithms which work correctly despite a limited number of failures without knowledge of their locations. As before, apart from this upper bound, we do not impose any restrictions on the values of false data.

The network misbehaviors would greatly jeopardize the performance of standard containment control algorithms, e.g., [18, 95, 92]. As one might imagine, if the normal agents are not equipped with any security strategies, even a single misbehaving node could be able to control the evolutions of normal states on its desire.

6.3.2 Resilient Containment Control

The above security concerns necessitate the study of resilient containment control. Let $x^i(k) \in \mathbb{R}$ be the state of agent i at kth instant. In this chapter, we focus on the stationary leaders. Particularly, denote $m^L \triangleq \min_{i \in \mathcal{L}_\mathcal{R}} x^i$ and $M^L \triangleq \max_{i \in \mathcal{L}_\mathcal{R}} x^i$ respectively be the minimum and maximum states of normal leaders. The resilient containment control aims to solve the following problem:

Resilient Containment Control Objective: For any initial states and network misbehaviors, it holds that

$$\lim_{k\to\infty} x^i(k) \in \delta, \ \forall i \in \mathcal{F}_{\mathcal{R}},$$

where $\delta \triangleq [m^L, M^L]$.

Therefore, through resilient protocols, the network could be avoided of being affected by the illegal behaviors too much. The purpose of this chapter is to determine the conditions under which normal followers resiliently move to δ, regardless of the actions of adversarial nodes. Both the first-order and second-order systems will be investigated in the following sections.

6.4 Resilient Containment Control of First-Order Systems

We first consider the first-order systems, where the dynamic for each agent is described by

$$x^i(k+1) = x^i(k) + u^i(k). \tag{6.1}$$

where $u^i(k)$ stands for the control input. While the misbehaving agents can perform arbitrarily, the normal ones should always follow the designed protocol. In particular, for the stationary leaders discussed, $u^i(k) = 0, \forall i \in \mathcal{L}_{\mathcal{R}}$. For the normal follower $i \in \mathcal{F}_{\mathcal{R}}$, it makes update as outlined in Algorithm 6.1.

Algorithm 6.1 Resilient containment control of first-order systems

1: Receive the states from all in-neighboring agents $j \in \mathcal{N}_i^+$. Form these values in a list $\mathcal{X}^i(k)$, and sort $\mathcal{X}^i(k)$ in an ascending order.
2: In $\mathcal{X}^i(k)$, remove f largest values that are higher than agent i's own state, i.e., $x^i(k)$. If there are less than f values higher than $x^i(k)$, then remove all of them.
3: Similar as that in Step 2, remove f smallest values that are lower than $x^i(k)$. If there are less than f values lower than $x^i(k)$, then remove all these values.
4: Denote $\mathcal{J}_i(k)$ as the set of agents whose values are retained at this time step. Agent i applies the control input as:

$$u^i(k) = \sum_{j\in\mathcal{J}_i(k)} d^i_j(k)(x^j(k) - x^i(k)), \tag{6.2}$$

such that $d^i_j(k) \geq \eta$ and $1 - \sum_{j\in\mathcal{J}_i(k)} d^i_j(k) \geq \eta$ for some $0 < \eta < 1$. Combining it with (6.1) yields the new state $x^i(k+1)$.
5: Transmit $x^i(k+1)$ to all out-neighbors $j \in \mathcal{N}_i^-$.

In view of Algorithm 6.1, it adopts a similar idea to that in most resilient protocols. At every step, the normal follower i obtains the states in its neighborhood, whereas up to f of them might be faulty. To ensure its state is updated safely, agent i needs to exclude the misleading information. Yet as it has no knowledge of the identities of these values, it ignores the most extreme ones, which are naturally referred to as be the largest or smallest numbers in this uni-dimensional system. In what follows, we shall theoretically prove the efficiency of such a protocol.

To proceed, we first introduce the following lemma:

Lemma 6.1 *Suppose the network $\mathcal{G} = (\mathcal{V}, \mathcal{E})$ is under f-local attack. If each normal follower $i \in \mathcal{F}_\mathcal{R}$ has at least $2f+1$ in-neighbors and makes update based on Algorithm 6.1, then there exists a nonempty set $\mathcal{M}_i^+(k) \subset \mathcal{N}_i^+ \cap \mathcal{R}$ and a set of weights $\{\bar{d}_j^i(k)\}$, such that the dynamic of agent i is mathematically equivalent to*

$$x^i(k+1) = \bar{d}_i^i(k)x^i(k) + \sum_{j \in \mathcal{M}_i^+(k)} \bar{d}_j^i(k)x^j(k). \tag{6.3}$$

Moreover, the following results hold:

a) Each weight in (6.3) is strictly positive, and $\bar{d}_i^i(k) + \sum_{j \in \mathcal{M}_i^+(k)} \bar{d}_j^i(k) = 1$;

b) For any normal agent in $\mathcal{J}_i(k)$, i.e., $w \in \mathcal{J}_i(k) \cap \mathcal{R}$, it holds that $w \in \mathcal{M}_i^+(k)$. Moreover, $\bar{d}_w^i(k) \geq \eta$.

c) $\bar{d}_i^i(k) \geq \eta$.

Proof *Since each normal follower has more than $2f+1$ in-neighbors, it is trivial to obtain that $\mathcal{J}_i(k) \neq \varnothing$.*

Clearly (6.1)-(6.2) can be expressed as

$$x^i(k+1) = d_i^i(k)x^i(k) + \sum_{j \in \mathcal{J}_i(k)} d_j^i(k)x^j(k), \tag{6.4}$$

where $d_i^i(k) = 1 - \sum_{j \in \mathcal{J}_i(k)} d_j^i(k) \geq \eta$. Thus every weight in the above equation is strictly positive. To show its equivalence to (6.3), we use the similar arguments as in Lemma 4.1. For the sake of completeness, we provide full proof here and consider the following cases:

- *$\mathcal{J}_i(k) \cap \mathcal{A} = \varnothing$, i.e., there is no adversarial agent in $\mathcal{J}_i(k)$.*
- *$\mathcal{J}_i(k) \cap \mathcal{A} \neq \varnothing$.*

As for the first case, the construction of $\bar{d}_j^i(k)$ is trivial by simply making it equal $d_j^i(k)$. Hence we focus on the second case where some misbehaving agents exist in $\mathcal{J}_i(k)$. Consider any agent $l \in \mathcal{J}_i(k) \cap \mathcal{A}$. Since $x^l(k)$ has been kept by agent i, it must be the case that either there are f neighboring values no less than $x^l(k)$, or agent i's own value $x^i(k)$ is not less than $x^l(k)$. Similarly, it is also true that either f neighboring values are not greater than $x^l(k)$, or

$x^i(k) \leq x^l(k)$. As there are at most f faulty ones in agent i's neighborhood, one could always find a pair of normal agents p, q such that $x^p(k) \leq x^l(k) \leq x^q(k)$. Hence, we have $x^l(k) = \gamma(k)x^p(k) + (1-\gamma(k))x^q(k)$ for some $0 \leq \gamma(k) \leq 1$. By setting $\bar{d}_p^i(k) = d_p^i(k) + \gamma(k)d_l^i(k)$ and $\bar{d}_q^i(k) = d_q^i(k) + (1-\gamma(k))d_l^i(k)$, the contribution of adversarial node l can be transformed to that of two normal ones, i.e., agents p and q. Hence one establishes the first part of this lemma. Claims b) and c) are then shown by noticing that for any $w \in \{i\} \cup (\mathcal{J}_i(k) \cap \mathcal{R})$, $\bar{d}_w^i(k) \geq d_w^i(k)$ holds.

Invoking Lemma 6.1, the misbehaving agents cannot have arbitrary control over the local states of normal agents. They can at most influence the "choice" of the convex combination weights $\bar{d}_j^i(k)$. As a result, the proposed strategy protects the normal agent from being significantly affected by the misbehaving ones.

With $\bar{d}_j^i(k)$ introduced in (6.3), let $\mathcal{G}(\bar{D}(k)) = (\mathcal{R}, \bar{\mathcal{E}}(k))$ be the *"virtual"* graph associated with $\bar{D}(k) = \{\bar{d}_j^i(k)\}$. We thus have the below result regarding $\mathcal{G}(\bar{D}(k)) = (\mathcal{R}, \bar{\mathcal{E}}(k))$.

Lemma 6.2 *Suppose the network $\mathcal{G} = (\mathcal{V}, \mathcal{E})$ is under f-local attack and is strongly $(3f+1)$-robust w.r.t $\mathcal{L}$. Consider the graph $\mathcal{G}(\bar{D}(k)) = (\mathcal{R}, \bar{\mathcal{E}}(k))$ induced by Algorithm 6.1. For any normal follower and at any time, there always exists at least one normal leader that has a directed path to it in $\mathcal{G}(\bar{D}(k))$.*

Proof *Consider the network $\mathcal{G} = (\mathcal{V}, \mathcal{E})$. Since it is strongly $(3f+1)$-robust w.r.t $\mathcal{L}$, there exists a nonempty set $\mathcal{S}_1 \subset \mathcal{F}_\mathcal{R} \subset \mathcal{V} \backslash \mathcal{L}$ such that for any $i \in \mathcal{S}_1$, it holds that $|\mathcal{N}_i^+ \backslash \mathcal{F}_\mathcal{R}| \geq 3f+1$. As $|\mathcal{N}_i^+ \cap \mathcal{A}| \leq f$ and $\mathcal{V} \backslash \mathcal{F}_\mathcal{R} = \mathcal{L} \cup \mathcal{A}$, one has $|\mathcal{N}_i^+ \cap \mathcal{L}_\mathcal{R}| \geq 2f+1$, which also implies that $\mathcal{L}_\mathcal{R} \neq \varnothing$. Namely, there always exist some normal leaders. According to Algorithm 6.1, agent i discards at most $2f$ neighboring values. Hence, at least one normal leader will be kept in $\mathcal{J}_i(k)$ at any time. Combining this with the second part of Lemma 6.1, agent i always has this normal leader as a in-neighbor in the induced graph $\mathcal{G}(\bar{D}(k))$.*

Next denote $\mathcal{F}_\mathcal{R}^1 \triangleq \mathcal{F}_\mathcal{R} \backslash \mathcal{S}_1 \subset \mathcal{V} \backslash \mathcal{L}$. If $\mathcal{F}_\mathcal{R}^1 \neq \varnothing$, one can construct another set $\mathcal{S}_2 \subset \mathcal{F}_\mathcal{R}^1$ containing all agents who has at least $3f+1$ neighbors outside $\mathcal{F}_\mathcal{R}^1$. Because the network is strongly $(3f+1)$-robust w.r.t $\mathcal{L}$, it holds that $\mathcal{S}_2 \neq \varnothing$. Then focus on any agent $j \in \mathcal{S}_2$. Since $|\mathcal{N}_j^+ \backslash \mathcal{F}_\mathcal{R}^1| \geq 3f+1$, as analyzed before, there always exists at least one normal agent in either $\mathcal{L}_\mathcal{R}$ or $\mathcal{S}_1$, labeled as m, whose value will be retained by agent j. Note that agent m either is a normal leader, or adopts a normal leader's state for updates. Recalling Lemma 6.1 again, there always exists a normal leader who has a directed path to agent j with length no more than 2 in $\mathcal{G}(\bar{D}(k))$.

Similarly, if $\mathcal{F}_\mathcal{R}^2 \triangleq \mathcal{F}_\mathcal{R}^1 \backslash \mathcal{S}_2 \neq \varnothing$, we define $\mathcal{S}_3 \subset \mathcal{F}_\mathcal{R}^2$ and collect all agents who have no less than $3f+1$ neighbors outside $\mathcal{F}_\mathcal{R}^2$ in $\mathcal{S}_3$. An analogical conclusion could then be derived that for each $l \in \mathcal{S}_3$, at least one agent in $\mathcal{L}_\mathcal{R} \cup \mathcal{S}_1 \cup \mathcal{S}_2$, whose value will be kept in $\mathcal{J}_l(k)$. Hence a "virtual" path exists from a benign leader to agent l.

Therefore, we can recursively define $\mathcal{F}_{\mathcal{R}}^{\tau} \triangleq \mathcal{F}_{\mathcal{R}}^{\tau-1} \backslash \mathcal{S}_{\tau}$. As long as $\mathcal{F}_{\mathcal{R}}^{\tau} \neq \varnothing$, let $\mathcal{S}_{\tau+1} \subset \mathcal{F}_{\mathcal{R}}^{\tau}$ be the collection of agents who has at least $3f+1$ neighbors outside $\mathcal{F}_{\mathcal{R}}^{\tau}$. Repeating the above analysis gives one similar results that any agent in $\mathcal{S}_{\tau+1}$ always retains some agent's value in $\mathcal{L}_{\mathcal{R}} \cup \mathcal{S}_1 \cup \mathcal{S}_2 \cup ... \cup \mathcal{S}_{\tau}$. Hence it can be reached by some normal leaders with the length no more than $\tau+1$. The proof thus completes.

Next let us recall the sequence of sets, namely $\{\mathcal{S}_i\}$, defined in the above proof. It is easy to conclude that some $n \leq |\mathcal{F}_{\mathcal{R}}|$ exists such that $\bigcap_{i=1}^{n} \mathcal{S}_i = \mathcal{F}_{\mathcal{R}}$.

With the above preparations, it is now ready to give the below result:

Theorem 6.1 *Suppose the network $\mathcal{G} = (\mathcal{V}, \mathcal{E})$ is under f-local attack and is strongly $(3f+1)$-robust w.r.t $\mathcal{L}$. Then the controller given by Algorithm 6.1 achieves the resilient containment control. Specifically, all normal followers will converge to δ regardless of the network misbehaviors.*

Proof *Firstly, we note that since the network is strongly $(3f+1)$-robust w.r.t $\mathcal{L}$, each agent $i \in \mathcal{F}_{\mathcal{R}}$ has at least $3f+1$ in-neighbors. To see this, suppose there exists a normal follower i with at most $3f$ in-neighbors. Choose $\mathcal{S}' \subseteq \mathcal{V} \backslash \mathcal{L} = \{i\}$ (defined in Definition 2.6). Then $\mathcal{S}'$ can be at most $3f$-reachable, implying that the graph is not strongly $(3f+1)$-robust w.r.t. $\mathcal{L}$. Since $f \geq 0$, the conditions of Lemma 6.1 hold.*

For notation convenience, let $m^F(k) \triangleq \min_{i \in \mathcal{F}_{\mathcal{R}}} x^i(k)$ and $M^F(k) \triangleq \max_{i \in \mathcal{F}_{\mathcal{R}}} x^i(k)$ respectively be the minimum and maximum value of normal followers' states at time k. In order to prove the state of each normal follower finally converge to the interval $\delta = [m^L, M^L]$, the below cases are considered:

1) *All normal followers are initially within δ, namely $\forall i \in \mathcal{F}_{\mathcal{R}}, x^i(0) \in \delta$;*
2) *There exists some $i \in \mathcal{F}_1 \subset \mathcal{F}_{\mathcal{R}}$ such that $x^i(0) > M^L$, while for any $j \in \mathcal{F}_{\mathcal{R}} \backslash \mathcal{F}_1$, $x^j(0) \in \delta$. That is, the initial states of some normal followers are greater than M^L, while the others are initially within δ;*
3) *There exists some $i \in \mathcal{F}_2 \subset \mathcal{F}_{\mathcal{R}}$ such that $x^i(0) < m^L$, while for any $j \in \mathcal{F}_{\mathcal{R}} \backslash \mathcal{F}_2$, $x^j(0) \in \delta$;*
4) *There exists some $i \in \mathcal{F}_3 \subset \mathcal{F}_{\mathcal{R}}$ such that $x^i(0) > M^L$, some $j \in \mathcal{F}_4 \subset \mathcal{F}_{\mathcal{R}}$ such that $x^j(0) < m^L$, and for other $l \in \mathcal{F}_{\mathcal{R}} \backslash (\mathcal{F}_3 \cup \mathcal{F}_4)$, $x^l(0) \in \delta$.*

Case 1): We shall prove by induction. Suppose at some time k, $x^i(k) \in \delta, \forall i \in \mathcal{F}_{\mathcal{R}}$. In view of (6.3), each normal follower updates its state based on a convex combination of some points in δ. Hence $x^i(k+1) \in \delta$, which completes our proof.

Case 2): In this case, we show that $M^F(k) \leq M^L$ as $k \to \infty$. The proof is comprised of two claims:

Claim 1: $M^F(k)$ is non-increasing. This can be easily obtained since at time $k+1$, each healthy follower updates as

$$x^i(k+1) = \bar{d}^i_i(k)x^i(k) + \sum_{j\in\mathcal{M}_i^+(k)} \bar{d}^i_j(k)x^j(k) \le M^F(k). \tag{6.5}$$

Claim 2: As long as $M^F(k) > M^L$, it will decrease after a finite time. To see this, let us consider the sequence of sets $\{\mathcal{S}_i\}_{i=1,2,\ldots,n}$. From former discussions, one knows every agent i in $\mathcal{S}_1$ retains at least one normal leader in $\mathcal{J}_i(k)$. Define $\gamma_0 = M^F(k) - M^L > 0$. Hence

$$\begin{aligned} x^i(k+1) &= \bar{d}^i_i(k)x^i(k) + \sum_{j\in\mathcal{M}_i^+(k)} \bar{d}^i_j(k)x^j(k) \\ &\le (1-\eta)M^F(k) + \eta(M^F(k)-\gamma_0) \\ &= M^F(k) - \eta\gamma_0, \end{aligned} \tag{6.6}$$

where the inequality holds due to Claim b) of Lemma 6.1 and we place the largest possible weight on $M^F(k)$. At $k+2$, the state of agent i is upper bounded by

$$\begin{aligned} x^i(k+2) &= \bar{d}^i_i(k+1)x^i(k+1) \\ &\quad + \sum_{j\in\mathcal{M}_i^+(k+1)} \bar{d}^i_j(k+1)x^j(k+1) \\ &\le \eta(M^F(k)-\eta\gamma_0) + (1-\eta)M^F(k+1) \\ &\le \eta(M^F(k)-\eta\gamma_0) + (1-\eta)M^F(k) \\ &= M^F(k) - \eta^2\gamma_0 \\ &< M^F(k), \end{aligned} \tag{6.7}$$

where the first inequality holds by invoking Claim c) of Lemma 6.1. Recursively, one concludes the state of any agent in $\mathcal{S}_1$ will always be smaller than $M^F(k)$ from $k+1$.

For the normal follower j in $\mathcal{S}_2$. As discussed before, agent j always has an in-neighbor either in $\mathcal{L}_\mathcal{R}$ or $\mathcal{S}_1$. Recalling (6.5) yields $x^j(k+1) \le M^F(k)$. At $k+2$, if one of agent j's in-neighbors belongs to $\mathcal{L}_\mathcal{R}$, the state of agent j is updated as

$$\begin{aligned} x^j(k+2) &= \bar{d}^j_j(k+1)x^j(k+1) + \sum_{l\in\mathcal{M}_j^+(k+1)} \bar{d}^j_l(k+1)x^l(k+1) \\ &\le (1-\eta)M^F(k) + \eta(M^F(k)-\gamma_0) \\ &= M^F(k) - \eta\gamma_0. \end{aligned} \tag{6.8}$$

On the other hand, if this in-neighbor is in $\mathcal{S}_1$, then the state is upper bounded by

$$\begin{aligned} x^j(k+2) &= \bar{d}^j_j(k+1)x^j(k+1) + \sum_{l\in\mathcal{M}_j^+(k+1)} \bar{d}^j_l(k+1)x^l(k+1) \\ &\leq (1-\eta)M^F(k) + \eta(M^F(k) - \eta\gamma_0) \\ &= M^F(k) - \eta^2\gamma_0, \end{aligned} \tag{6.9}$$

where the inequality is due to the fact that agent j has a direct neighbor in $\mathcal{S}_1$ whose state is upper bounded by (6.6) and whose weight is lower bounded by η (by invoking Lemma 6.1.b)). Combining the above two relations gives that $x^j(k+2) \leq M^F(k) - \eta^2\gamma_0$. As before, one knows for any agent in $\mathcal{S}_2$, its value will be less than $M^F(k)$ from $k+2$.

Similarly, one can repeat the above analysis and conclude for any agent in $\mathcal{S}_\tau$, its state will definitely decrease below $M^F(k)$ after τ steps. Hence, from $k+n$, every normal follower has its state strictly less than $M^F(k)$. Namely $M^F(k+n) < M^F(k)$. Since $n \leq |\mathcal{F}_\mathcal{R}|$, thus the proof of Step 2 completes.

In view of Claims 1 and 2, we know that $M^F(k) \leq M^L$ as $k \to \infty$. Based on (6.3), note also that $m^F(k) \leq m^L$ at any $k > 0$. Hence all normal followers finally converge to δ.

Case 3): The analysis of this case is similar to that of Case 2). One knows as k goes to infinity, the state of any normal follower will converge to δ.

Case 4): Combining Cases 2) and 3), the proof completes.

6.5 Resilient Containment Control of Second-Order Systems

Given that a broad class of autonomous agents (e.g., vehicles, sensors) requires a double-integrator model (see, for examples, [16, 86] and [39]), this section discusses the resilient containment control in second-order systems. In particular, each follower $i \in \mathcal{F}$ is governed by both position and velocity states as below:

$$\begin{aligned} x^i(k+1) &= x^i(k) + v^i(k), \\ v^i(k+1) &= v^i(k) + u^i(k), \end{aligned} \tag{6.10}$$

where $u^i(k) \in \mathbb{R}$ is the control signal. For the normal follower $i \in \mathcal{F}_\mathcal{R}$, it follows the updating rule as stated in Algorithm 6.2.

Remark 6.1 *Note that ρ is introduced in the control law (6.11) to stabilize the double-integrator dynamics. Furthermore, $v^j(k) = 0$ for any $j \in \mathcal{L}_\mathcal{R}$.*

Algorithm 6.2 Resilient containment control of second-order systems

1-3: Steps 1 to 3 are the same as those in Algorithm 6.1.
4: Denote $\mathcal{J}_i(k)$ as the set of agents whose values are retained at this time step. Agent i applies the control input as:

$$u^i(k) = -2\rho v^i(k) + \sum_{j \in \mathcal{J}_i(k)} d^i_j(k)(x^j(k) - x^i(k)). \tag{6.11}$$

In the above equation, $\rho > 0$, each weight $d^i_j(k) \geq \eta$, and $\sum_{j \in \mathcal{J}_i(k)} d^i_j(k) \leq 1 - \eta$ for some $0 < \eta < 1$. Combining it with (6.10) yields the new state $x^i(k+1)$.
5: Transmit $x^i(k+1)$ to all out-neighbors $j \in \mathcal{N}_i^-$.

To show the resiliency of Algorithm 6.2, Lemma 6.3 would be needed. The proof is established in a similar manner to that of Lemma 6.1 and hence omitted.

Lemma 6.3 *Suppose the network $\mathcal{G} = (\mathcal{V}, \mathcal{E})$ is under f-local attack. If each normal follower $i \in \mathcal{F}_\mathcal{R}$ has at least $2f+1$ in-neighbors and makes update based on Algorithm 6.2, then there exists a nonempty set $\mathcal{M}_i^+(k) \subset \mathcal{N}_i^+ \cap \mathcal{R}$ and a set of weights $\{\bar{d}^i_j(k)\}$, such that its control law (6.11) is mathematically equivalent to*

$$u^i(k) = -2\rho v^i(k) + \sum_{j \in \mathcal{M}_i^+(k)} \bar{d}^i_j(k)(x^j(k) - x^i(k)), \tag{6.12}$$

where each weight is strictly positive and $\sum_{j \in \mathcal{M}_i^+(k)} \bar{d}^i_j(k) \leq \sum_{j \in \mathcal{J}_i(k)} d^i_j(k) \leq 1 - \eta$. Furthermore, for any $w \in \mathcal{J}_i(k) \cap \mathcal{R}$, it also holds that $w \in \mathcal{M}_i^+(k)$ and $\bar{d}^i_w(k) \geq \eta$.

As analyzed before, Algorithm 6.2 equivalently leads to a network $\mathcal{G}(\bar{D}(k)) = (\mathcal{R}, \bar{\mathcal{E}}(k))$ associated with $\bar{D}(k) = \{\bar{d}^i_j(k)\}$. Furthermore, applying the same argument as that in the proof of Lemma 6.2, one obtains the below result:

Lemma 6.4 *If $\mathcal{G} = (\mathcal{V}, \mathcal{E})$ is under f-local attack and is strongly $(3f+1)$-robust w.r.t $\mathcal{L}$, then for any normal follower and at any time, there always exists at least one normal leader that has a directed path to it in $\mathcal{G}(\bar{D}(k)) = (\mathcal{R}, \bar{\mathcal{E}}(k))$.*

Before moving on, we also need the following lemma:

Lemma 6.5 ([79]) *Let $\mathcal{Y}$ be a nonempty closed convex set in $\mathbb{R}^d$. For any $y^i \in \mathbb{R}^d$, it holds that*

$$\left\| \sum_{i=1}^{l} a^i y^i - \mathcal{P}_\mathcal{Y}\Big(\sum_{i=1}^{l} a^i y^i\Big) \right\| \leq \sum_{i=1}^{l} a^i \| y^i - \mathcal{P}_\mathcal{Y}(y^i) \|,$$

where each $a^i \geq 0$ and $\sum_{i=1}^{l} a^i = 1$.

Now we are in position of presenting the below result:

Theorem 6.2 *Suppose the network $\mathcal{G} = (\mathcal{V}, \mathcal{E})$ is under f-local attack and is strongly $(3f+1)$-robust w.r.t $\mathcal{L}$. If $\sqrt{1-\eta} < \rho < 1$, then the controller given by Algorithm 6.2 achieves the resilient containment control. Specifically, all normal followers will converge to δ regardless of the network misbehaviors.*

Proof *As analyzed in the proof of Theorem 6.1, the conditions of Lemma 6.3 hold.*

Then motivated by that in [147], let us introduce an auxiliary variable as $\phi^i(k) \triangleq x^i(k) + v^i(k)/\rho$. Hence $\phi^i(k) = x^i(k), \forall i \in \mathcal{L}_{\mathcal{R}}$. Now from (6.10) *and* (6.12)*, one has the below relations for any $i \in \mathcal{F}_{\mathcal{R}}$:*

$$
\begin{aligned}
&x^i(k+1) = (1-\rho)x^i(k) + \rho\phi^i(k), \\
&\phi^i(k+1) = (1-\rho)\phi^i(k) + \Big(\rho - 1/\rho \sum_{j\in\mathcal{M}_i^+(k)} \bar{d}_j^i(k)\Big)x^i(k) \\
&\qquad\qquad + 1/\rho \sum_{j\in\mathcal{M}_i^+(k)} \bar{d}_j^i(k)x^j(k).
\end{aligned}
\tag{6.13}
$$

Then define the following Lyapunov function:

$$
V(k) \triangleq \max_{i\in\mathcal{F}_{\mathcal{R}}} \big(\max(|x^i(k) - \mathcal{P}_\delta(x^i(k))|, |\phi^i(k) - \mathcal{P}_\delta(\phi^i(k))|)\big). \tag{6.14}
$$

Namely $V(k)$ is the maximum of all the distance from $x^i(k)$ and $\phi^i(k)$ to the set δ among normal followers. Also note that for any $j \in \mathcal{L}_{\mathcal{R}}$, $|x^j(k) - \mathcal{P}_\delta(x^j(k))| = 0$ and $|\phi^j(k) - \mathcal{P}_\delta(\phi^j(k))| = 0$.

At time $k+1$, consider any $i \in \mathcal{F}_{\mathcal{R}}$. In view of (6.13) *and Lemma 6.5, it holds that*

$$
\begin{aligned}
&|x^i(k+1) - \mathcal{P}_\delta(x^i(k+1))| \\
\le&(1-\rho)|x^i(k) - \mathcal{P}_\delta(x^i(k))| + \rho|\phi^i(k) - \mathcal{P}_\delta(\phi^i(k))| \\
\le&V(k).
\end{aligned}
\tag{6.15}
$$

On the other hand, given that $\sum_{j\in\mathcal{M}_i^+(k)} \bar{d}_j^i(k) \le 1-\eta < \rho^2 < 1$, one has $0 < \rho - 1/\rho\sum_{j\in\mathcal{M}_i^+(k)} \bar{d}_j^i(k) < 1$. Hence

$$
\begin{aligned}
&|\phi^i(k+1) - \mathcal{P}_\delta(\phi^i(k+1))| \\
\le&(1-\rho)|\phi^i(k) - \mathcal{P}_\delta(\phi^i(k))| + \Big(\rho - 1/\rho \sum_{j\in\mathcal{M}_i^+(k)} \bar{d}_j^i(k)\Big)|x^i(k) - \mathcal{P}_\delta(x^i(k))| \\
&+ 1/\rho \sum_{j\in\mathcal{M}_i^+(k)} \bar{d}_j^i(k)|x^j(k) - \mathcal{P}_\delta(x^j(k))| \\
\le&\Big(1 - 1/\rho \sum_{j\in\mathcal{M}_i^+(k)\cap\mathcal{L}_{\mathcal{R}}} \bar{d}_j^i(k)\Big)V(k) \\
\le&V(k).
\end{aligned}
\tag{6.16}
$$

As (6.15) and (6.16) hold for any $i \in \mathcal{F}_\mathcal{R}$, $V(k+1) \le V(k)$. Next we shall study the limit of $V(k)$. To this end, we make the following claims:

Claim 1): For any $i \in \mathcal{F}_\mathcal{R}$, if $|x^i(t) - \mathcal{P}_\delta(x^i(t)| \le \epsilon_1 V(k)$ for some $t \ge k$ and $0 \le \epsilon_1 < 1$, then there exists $0 \le \epsilon_2 < 1$, such that $|x^i(t+1) - \mathcal{P}_\delta(x^i(t+1)| \le \epsilon_2 V(k)$ holds.

To see this, let us recall (6.15) again. One thus has

$$\begin{aligned}
&|x^i(t+1) - \mathcal{P}_\delta(x^i(t+1))| \\
\le&(1-\rho)|x^i(t) - \mathcal{P}_\delta(x^i(t))| + \rho|\phi^i(t) - \mathcal{P}_\delta(\phi^i(t))| \\
\le&(1-\rho)\epsilon_1 V(k) + \rho V(t) \\
\le&(1-\rho)\epsilon_1 V(k) + \rho V(k) \\
=&((1-\rho)\epsilon_1 + \rho)V(k).
\end{aligned} \tag{6.17}$$

Claim 1) is concluded by letting $\epsilon_2 = (1-\rho)\epsilon_1 + \rho$.

Claim 2): For any $i \in \mathcal{F}_\mathcal{R}$, if there exists an agent in $\mathcal{J}_i(k) \cap \mathcal{R}$, labeled as l, such that $|x^l(t) - \mathcal{P}_\delta(x^l(t)| \le \epsilon_3 V(k)$ for some $t \ge k$ and $0 \le \epsilon_3 < 1$, then there exists $0 \le \epsilon_4 < 1$ such that $|\phi^i(t+1) - \mathcal{P}_\delta(\phi^i(t+1)| \le \epsilon_4 V(k)$.

Recall Lemma 6.4 and Eqn. (6.16), we have

$$\begin{aligned}
&|\phi^i(t+1) - \mathcal{P}_\delta(\phi^i(t+1))| \\
\le&(1-\rho)|\phi^i(t) - \mathcal{P}_\delta(\phi^i(t))| + \Big(\rho - 1/\rho \sum_{j \in \mathcal{M}_i^+(t)} \bar{d}_j^i(t)\Big)|x^i(t) - \mathcal{P}_\delta(x^i(t))| \\
&+ 1/\rho \sum_{j \in \mathcal{M}_i^+(t)} \bar{d}_j^i(t)|x^j(t) - \mathcal{P}_\delta(x^j(t))| \\
\le&\Big(1 - 1/\rho \sum_{j \in \mathcal{M}_i^+(t)} \bar{d}_j^i(t)\Big)V(t) + 1/\rho \sum_{j \in \mathcal{M}_i^+(t)\setminus\{l\}} \bar{d}_j^i(t)V(t) + \frac{\epsilon_3}{\rho}\bar{d}_l^i(t)V(k) \\
\le&\Big(1 - \frac{1-\epsilon_3}{\rho}\bar{d}_l^i(t)\Big)V(k) \\
\le&\Big(1 - \frac{1-\epsilon_3}{\rho}\eta\Big)V(k).
\end{aligned} \tag{6.18}$$

Setting $\epsilon_4 = 1 - (1-\epsilon_3)\eta/\rho$ yields Claim 2).

Claim 3): For any $i \in \mathcal{F}_\mathcal{R}$, if $|\phi^i(t) - \mathcal{P}_\delta(\phi^i(t)| \le \epsilon_5 V(k)$ for some $t \ge k$ and $0 \le \epsilon_5 < 1$, then there exists $0 \le \epsilon_6 < 1$ such that $|x^i(t+1) - \mathcal{P}_\delta(x^i(t+1)| \le \epsilon_6 V(k)$.

This claim is made from (6.15). That is,

$$\begin{aligned}
&|x^i(t+1) - \mathcal{P}_\delta(x^i(t+1))| \\
\le&(1-\rho)|x^i(t) - \mathcal{P}_\delta(x^i(t))| + \rho|\phi^i(t) - \mathcal{P}_\delta(\phi^i(t))| \\
\le&(1-\rho)V(k) + \rho\epsilon_5 V(k).
\end{aligned} \tag{6.19}$$

By letting $\epsilon_6 = 1 - \rho(1-\epsilon_5)$, the proof of this claim completes.

Finally let us consider the set $\mathcal{S}_1$ defined in the proof of Lemma 6.2. For any $i \in \mathcal{S}_1$, it has some normal leaders in its in-neighborhood. Then from (6.16), $|\phi^i(k+1) - \mathcal{P}_\delta(\phi^i(k+1))| < V(k)$ *holds by noticing that* $\sum_{j \in \mathcal{M}_i^+(k) \cap \mathcal{L}_\mathcal{R}} \bar{d}_j^i(k) > 0$*. Based on Claims 1)–3) and Lemmas 6.3–6.4, one knows* $\lim_{k\to\infty} V(k) = 0$ *and thus the proof of this theorem finishes.*

In view of Theorems 6.1 and 6.2, the convergence of the proposed algorithms does not depend on the actions of misbehaving agents. Hence they work resiliently even in the worst-case scenario, where the misbehaving agents could have full knowledge of graph topology, updating rules, etc, and could also be able to send different information to different out-neighbors. Therefore, Algorithms 6.1 and 6.2 can be safely used under any assumptions involving faults or attacks, as long as no more than f misbehaving nodes exist in the normal follower's in-neighborhood.

Remark 6.2 *As indicated in Theorems 6.1 and 6.2, the maximum number of allowable misbehaving nodes depends directly on the communication topology. Particularly, the network should be strongly* $(3f+1)$*-robust w.r.t* $\mathcal{L}$*, to ensure that each normal follower is capable of tolerating at most* f *attacks in its in-neighborhood. That is, the network needs to be "connected" enough to increase its resilience. This also indicates a trade-off between reducing communication burden and system security.*

Remark 6.3 *Invoking Lemmas 6.1 and 6.3, both resilient algorithms result in the control law being equivalent to the containment control under a switching topology. Therefore, in such cases, the final states of all normal followers might not be constant, although remain in the convex hull spanned by the stationary leaders [18].*

6.6 Numerical Example

In this section, we provide some numerical examples to illustrate the proposed algorithms and verify the theoretical results established before.

We consider the communication network given by Figure 6.1, in which $\mathcal{L} = \{1, 2, 3, 4\}$ and $\mathcal{F} = \{5, 6, 7, 8\}$. Clearly the graph is strongly 4-robust w.r.t. $\mathcal{L}$. Theorems 6.1 and 6.2 indicate that the network should be able tolerate a single misbehaving node. To verify this, we consider two examples.

In our first example, first-order systems are considered. Suppose the leader agent 1 is compromised. It intends to followers from getting into the desired interval by setting its states as $x^1(k) = 10 * \sin(k/5) + 2$ at any $k \geq 0$. On the other hand, the normal agents are initialized with $x^2(0) = 3, x^3(0) = 2.5, x^4(0) = 1.2, x^5(0) = -2, x^6(0) = 6, x^7(0) = 2, x^8(0) = 3.2$, and normal followers always make updates based on the predefined algorithms.

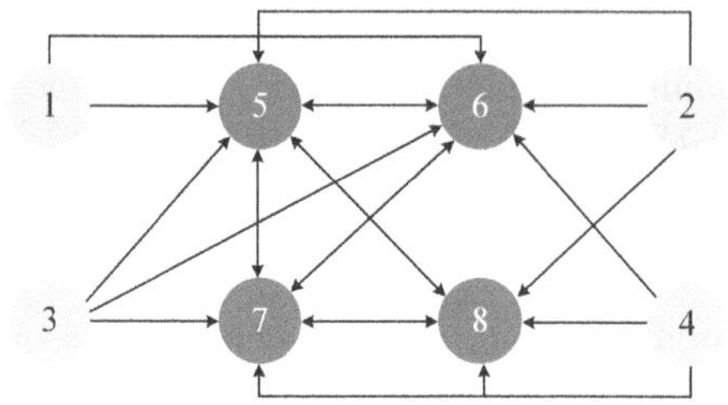

FIGURE 6.1: Communication topology.

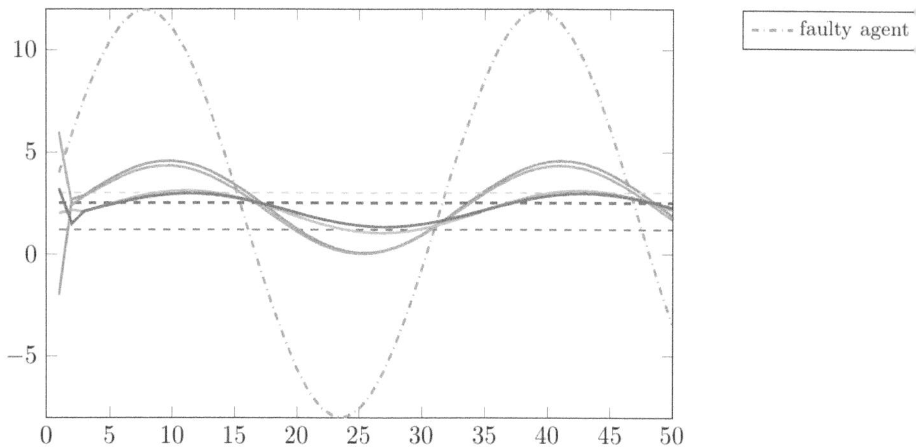

FIGURE 6.2: Local states of each agent according to the containment control algorithm in [17], where the dashed and solid lines represent the states of normal leaders and followers, respectively.[2]

Firstly, Figure 6.2 presents the performance of the traditional containment control algorithm ([17]) in this adversarial environment. The results show that the normal followers would be seriously affected by the misbehaviors and move out of the convex hull formed by normal leaders, necessitating resilient controllers.

As a comparison, we next illustrate the performance of Algorithm 6.1. Set $\eta = 0.1$. For simplicity, let the updating weights be $d_j^i(k) = (|\mathcal{J}^i(k)| + 1)^{-1}$ for each $j \in \mathcal{J}^i(k)$, which satisfies that $d_j^i(k) \geq \eta$ and $1 - \sum_{j\in\mathcal{J}_i(k)} d_j^i(k) \geq \eta$. The result is depicted in Figure 6.3, showing that normal followers ultimately move into the convex hull spanned by the normal leaders, which validates Theorem 6.1.

As another example, we investigate the scenario for second-order systems. In this case, follower 5 is misbehaving, which randomly sets its state within

[2]Reproduced with permission of ©2013 IEEE.

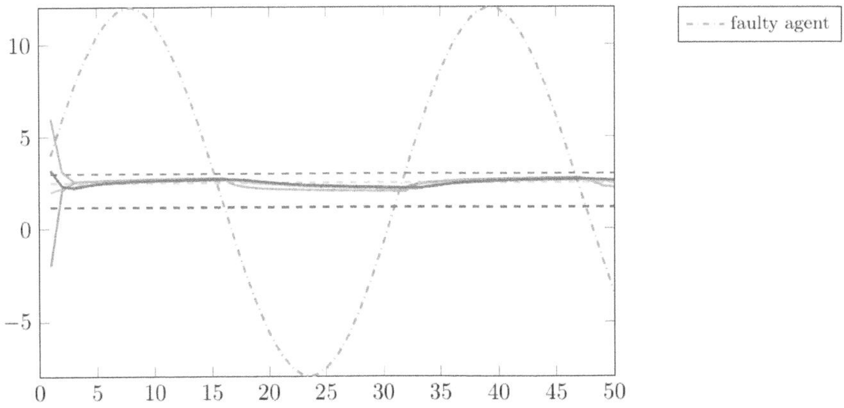

FIGURE 6.3: Local states of each agent along iteration in first-order system, where the dashed and solid lines represent the states of normal leaders and followers, respectively.[3]

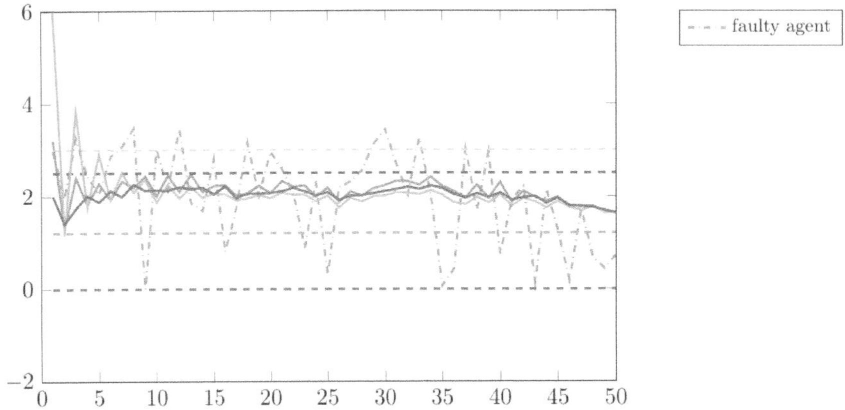

FIGURE 6.4: Local states of each agent along iteration in second-order system, where the dashed and solid lines represent the states of normal leaders and followers, respectively.[4]

the range $[0, 3.5]$. ρ is set to be 0.75 in Algorithm 6.2, satisfying that $\sqrt{1-\eta} < \rho < 1$. Despite the misbehaviors, results in Figure 6.4 verify the effectiveness of the proposed strategy.

[3]Reproduced with permission of ©2013 IEEE.
[4]Reproduced with permission of ©2013 IEEE.

6.7 Conclusion

In this chapter, we consider resilient containment control with stationary leaders. Despite some misbehaving agents, secure protocols are desired to drive the normal followers into the convex hull spanned by the normal leaders. Towards this goal, two resilient algorithms respectively for first-order and second-order systems are presented. It is proved that both algorithms are resilient to the f-local attack as long as the network is strongly $3f+1$ robust w.r.t $\mathcal{L}$.

Future work involves the extension of our results to the more general high-dimensional spaces, to guarantee that all normal followers converge to the convex hull spanned by normal leaders. As mentioned in the introduction, the main idea in resilient protocols is to discard the most extreme values in the normal agents' neighborhoods. However, as the vector space is equipped with a partial order, how to define "extreme" is one of the most challenges in resilient multi-dimensional algorithms. Furthermore, topological conditions should be developed under which each normal follower could be reached by at least one normal leader in the "virtual" network. Alternatively, one can also simply apply Algorithms 6.1 and 6.2 to each entry of the vector state. This approach guarantees the normal followers converge to the minimum hyperrectangle containing normal leaders' states and hence protects them from being arbitrarily affected. Although the normal followers may fail to converge to the convex hull formed by these states, this approach secures the system to some extent.

7

Conclusions and Future Work

This chapter concludes the whole book by summarizing its contents and highlighting its major contributions. Moreover, some future research perspectives with regard to CPS security will be discussed.

7.1 Conclusions

In this book, resilient detection and control in CPSs is discussed, which is particularly motivated by the security concerns of deception attacks on communication channels. Throughout this book, we consider the attack model where the number of compromised sensors/agents is upper bounded. Note this upper bound might be determined by the *a priori* knowledge about the quality of sensors. It can also be viewed as a design parameter, which indicates the resilience level that the system is willing to introduce. Yet, despite this upper bound, we do not pose any assumptions on the patterns or values of the malicious data introduced by the adversary. Namely, the compromised components are allowed to transmit arbitrary data. They could also collude among themselves to decide on the deceptive values to be communicated. On the other hand, the system operator or benign agents only know this upper bound but have no information on the identities of others. Working against this attack model, the following problems have been addressed in an the book:

Driven by the fact that existing distributed algorithms may fail in an adversarial environment with unreliable communication channels, the resilient coordination of MASs is discussed in Chapters 4–6. Specifically, a continuous-time second order MAS is considered in Chapter 4. Under malicious nodes, the benign ones still aim to achieve an agreement among themselves. Towards this end, an impulsive consensus algorithm is proposed, which, by using position states only, facilitates the resilient consensus among benign nodes while avoiding the continuous transmission and control actions.

In many applications of MASs, the agents not only hope to achieve an agreement, but also want this value "fairly" represents their initial states. As the average consensus can never be achieved in an adversarial environment, Chapter 5 seeks for a resilient algorithm which facilitates the agreement within the convex hull formed by the benign agents' initial states. Yet, when dealing

DOI: 10.1201/9781003409199-7

with high-dimensional systems, the computation cost increases inevitably with the system dimensionality. To address this issue, an efficient approach is also proposed to relieve this burden through linear programming. Compared with the existing solutions, our approach introduces lower complexity.

Besides the leaderless consensus, the resilient containment control with multiple leaders is finally investigated in Chapter 6. Regardless of the network misbehavior, resilient controllers are designed to drive the benign followers to move into the convex hull formed by benign leaders. Both the first-order and second-order systems are considered, where convexity analysis and the Lyapunov approach are respectively adopted to validate the efficiency of the proposed algorithms.

7.2 Future Work

On the basis of the aforementioned research works, this section further comes up with some new perspectives on the security and resilience of cyber-physical systems. They are briefly summarized as follows.

7.2.1 Secure Coordination in MASs

In this book, some results on resilient consensus has been obtained. Since the consensus among agents serves as the basic objective in distributed coordination, this book provides tools to increase the resiliency of other consensus-based formulations as well. For example, in distributed optimization, the agents aim to agree on the minimizer of a global objective function. As proved in [130], any distributed algorithm that is guaranteed to output a globally optimum value in the absence of adversaries, can be arbitrarily co-opted by network misbehaviors. This means there exists a trade-off between the optimality and security. To be specific, by combining the solutions in distributed optimization and resilient consensus, a secure distributed optimization algorithm can be proposed, which protects the benign agents from being seriously affected by the misbehaviors. On the other hand, given the aforementioned trade-off, we have to sacrifice the optimality of this algorithm during normal operations with an aim to increase its security. That is, the resilient strategy will lead to a "sub-optimal" solution in the absence of misbehaviors. In this case, a characterization of the "distance-to-optimality" will be studied. Other examples include distributed estimation in sensor networks, formation control of multi-robot systems, etc.

Another interesting topic is to analyze the security in MASs with game-theoretic tools. For example, the distributed optimal control is to minimize the cost function of a network in a dynamical process with an optimal strategy. The attacker, on the other hand, intends to maximize this cost by injecting malicious data to the update of agents. This problem may be formulated as a zero-sum game and addressed by mathematical tools in that field.

7.2.2 Applications to the Secure Coordination of More Complicated Cyber-Physical Systems

We are expecting the theoretical results in this book can be applied to enhance the security of practical systems. Yet, we should note that the theoretical works focus more on the level of control and optimization while ignoring the physical constraints. For example, one problem that will be encountered in the physical implementation is that MASs are usually composed of autonomous components that have limited communication capabilities. Therefore, it is important to equip the control algorithms with robust mechanisms such that they could account for the errors caused by the communication limitations like sensing noises, time delays, quantization, etc. The conventional methodologies to address this issue include the introduction of quantizers and the implementation of asynchronous communication actions. Moreover, since dedicated hardware can only operate at some maximum frequency (e.g., a physical device can only broadcast a message or evaluate a function for a finite number of times in any finite period of time), the event/self-triggered sampling and control can contribute to the physical implementation. However, the aforementioned strategies need to be designed very carefully in the presence of malicious and unexpected behaviors, as the cunning attackers can take advantage of the asynchrony in communication to prevent non-faulty components from reaching control objectives [34]. A possible solution to enhance the system resiliency is to introduce randomization in the control rules so that the adversary cannot predict the update times in advance.

What's more, in this book, we assume the local agents only interact through communication networks. Nevertheless, in practical systems, there might exist physical couplings among agents. A prospective aspect lies in the power system operation, where multiple generation units cooperatively achieve primary targets, such as frequency/voltage regulations, load power sharing, and economic dispatch. These generators are connected through electrical wires in the physical layer and thus electrically coupled. To make these targets immune from malicious cyber attacks, resilient control schemes are required to better coordinate multiple components, where physical constraints, such as power flow equations [67], must be taken into account.

7.2.3 Privacy Preserving in Networked Control Systems

As the distributed control systems become more widespread, concerns are growing about how these systems collect and make use of the privacy-sensitive data obtained from participating individuals. In many applications, these individuals may not want to disclose their real-time information when cooperating with others to complete a global task. For example, in social networks, a group of participants would like to employ a consensus algorithm to achieve a common opinion on a subject while keeping their personal comments on this subject secret. Hence, data privacy has become one of the most emerging topics in CPSs. To protect it, one common approach is differential privacy, which limits the disclosure of private information by introducing randomness into the protected data. In general, there exists a trade-off between data privacy and utility. Hence, the coming problems are how to design the random variables and how to guarantee the control and optimization performance under this randomness.

Bibliography

[1] Comprehensive report on ukraine power system attacks. Website, 2016. https://www.antiy.net/p/comprehensive-analysis-report-on-ukraine-power-system-attacks/.

[2] Our choice of digital signature algorithm. Website, 2017. https://exonum.com/blog/09-27-17-digital-signature/.

[3] Waseem Abbas, Aron Laszka, and Xenofon Koutsoukos. Improving network connectivity and robustness using trusted nodes with application to resilient consensus. *IEEE Transactions on Control of Network Systems*, 5(4):2036–2048, 2017.

[4] Waseem Abbas, Yevgeniy Vorobeychik, and Xenofon Koutsoukos. Resilient consensus protocol in the presence of trusted nodes. In *2014 7th International Symposium on Resilient Control Systems (ISRCS)*, pages 1–7. IEEE, 2014.

[5] A. Abrardo, M. Barni, K. Kallas, and B. Tondi. A game-theoretic framework for optimum decision fusion in the presence of byzantines. *IEEE Transactions on Information Forensics and Security*, 11(6):1333–1345, June 2016.

[6] Pankaj K Agarwal, Micha Sharir, and Emo Welzl. Algorithms for center and tverberg points. *ACM Transactions on Algorithms (TALG)*, 5(1):1–20, 2008.

[7] Hossein Ahmadi, Tarek Abdelzaher, Jiawei Han, Nam Pham, and Raghu K Ganti. The sparse regression cube: A reliable modeling technique for open cyber-physical systems. In *Proceedings of the 2011 IEEE/ACM Second International Conference on Cyber-Physical Systems*, pages 87–96. IEEE Computer Society, 2011.

[8] Radhakisan Baheti and Helen Gill. Cyber-physical systems. *The impact of control technology*, 12:161–166, 2011.

[9] Suat Bayram and Sinan Gezici. On the restricted neyman–pearson approach for composite hypothesis-testing in presence of prior distribution uncertainty. *IEEE Transactions on Signal Processing*, 59(10):5056–5065, 2011.

[10] James O Berger. Statistical decision theory and bayesian analysis, 1985.

[11] Daniel J Bernstein. Cost analysis of hash collisions: Will quantum computers make sharcs obsolete. *SHARCS*, 9:105, 2009.

[12] Debnath Bhattacharyya, Rahul Ranjan, Farkhod Alisherov, Minkyu Choi, et al. Biometric authentication: A review. *International Journal of u-and e-Service, Science and Technology*, 2(3):13–28, 2009.

[13] Stephen Boyd, Stephen P Boyd, and Lieven Vandenberghe. *Convex optimization.* Cambridge university press, 2004.

[14] Davide Calvaresi, Kevin Appoggetti, Luca Lustrissimini, Mauro Marinoni, Paolo Sernani, Aldo Franco Dragoni, and Michael Schumacher. Multi-agent systems' negotiation protocols for cyber-physical systems: Results from a systematic literature review. In *ICAART (1)*, pages 224–235, 2018.

[15] Ming Cao, A Stephen Morse, and Brian DO Anderson. Reaching a consensus in a dynamically changing environment: A graphical approach. *SIAM Journal on Control and Optimization*, 47(2):575–600, 2008.

[16] Y. Cao, D. Stuart, W. Ren, and Z. Meng. Distributed containment control for multiple autonomous vehicles with double-integrator dynamics: Algorithms and experiments. *IEEE Transactions on Control Systems Technology*, 19(4):929–938, July 2011.

[17] Yongcan Cao and Wei Ren. Containment control with multiple stationary or dynamic leaders under a directed interaction graph. In *Proceedings of the 48h IEEE Conference on Decision and Control (CDC) held jointly with 2009 28th Chinese Control Conference*, pages 3014–3019. IEEE, 2009.

[18] Yongcan Cao, Wei Ren, and Magnus Egerstedt. Distributed containment control with multiple stationary or dynamic leaders in fixed and switching directed networks. *Automatica*, 48(8):1586–1597, 2012.

[19] Alvaro A Cárdenas, Saurabh Amin, and Shankar Sastry. Research challenges for the security of control systems. In *HotSec*, 2008.

[20] Alvaro A Cardenas, Saurabh Amin, and Shankar Sastry. Secure control: Towards survivable cyber-physical systems. In *Distributed Computing Systems Workshops, 2008. ICDCS'08. 28th International Conference on*, pages 495–500. IEEE, 2008.

[21] Rohan Chabukswar and Bruno Sinopoli. Secure detection with correlated binary sensors. In *American Control Conference (ACC), 2015*, pages 3874–3879. IEEE, 2015.

[22] Jian Chen and Ali Abur. Placement of pmus to enable bad data detection in state estimation. *IEEE Transactions on Power Systems*, 21(4):1608–1615, 2006.

[23] Jie Chen and Ron J Patton. *Robust model-based fault diagnosis for dynamic systems*, volume 3. Springer Science & Business Media, 2012.

[24] Herman Chernoff. A measure of asymptotic efficiency for tests of a hypothesis based on the sum of observations. *The Annals of Mathematical Statistics*, pages 493–507, 1952.

[25] Jorge Cortés, Geir E Dullerud, Shuo Han, Jerome Le Ny, Sayan Mitra, and George J Pappas. Differential privacy in control and network systems. In *2016 IEEE 55th Conference on Decision and Control (CDC)*, pages 4252–4272. IEEE, 2016.

[26] Gyorgy Dan and Henrik Sandberg. Stealth attacks and protection schemes for state estimators in power systems. In *2010 first IEEE international conference on smart grid communications*, pages 214–219. IEEE, 2010.

[27] Ludwig Danzer, Branko Grünbaum, and Victor Klee. Helly's theorem and its relatives. *Proceedings of Symposia in Pure Mathematics*, pages 101–180, 1963.

[28] Derek L Davis and Lionel Smith. Authentication system based on periodic challenge/response protocol, July 11 2000. US Patent 6,088,450.

[29] Marcelo M de Azevedo and Douglas M. Blough. Multistep interactive convergence: An efficient approach to the fault-tolerant clock synchronization of large multicomputers. *IEEE Transactions on Parallel and Distributed Systems*, 9(12):1195–1212, 1998.

[30] Ruilong Deng, Gaoxi Xiao, Rongxing Lu, Hao Liang, and Athanasios V Vasilakos. False data injection on state estimation in power systems—attacks, impacts, and defense: A survey. *IEEE Transactions on Industrial Informatics*, 13(2):411–423, 2017.

[31] Dorothy E Denning and Giovanni Maria Sacco. Timestamps in key distribution protocols. *Communications of the ACM*, 24(8):533–536, 1981.

[32] Seyed Mehran Dibaji and Hideaki Ishii. Consensus of second-order multi-agent systems in the presence of locally bounded faults. *Systems & Control Letters*, 79:23–29, 2015.

[33] Seyed Mehran Dibaji and Hideaki Ishii. Resilient multi-agent consensus with asynchrony and delayed information. *IFAC-PapersOnLine*, 48(22):28–33, 2015.

[34] Seyed Mehran Dibaji, Hideaki Ishii, and Roberto Tempo. Resilient randomized quantized consensus. *IEEE Transactions on Automatic Control*, 63(8):2508–2522, 2017.

[35] Seyed Mehran Dibaji, Mohammad Pirani, David Bezalel Flamholz, Anuradha M Annaswamy, Karl Henrik Johansson, and Aranya Chakrabortty. A systems and control perspective of cps security. *Annual Reviews in Control*, 2019.

[36] Kemi Ding, Subhrakanti Dey, Daniel E Quevedo, and Ling Shi. Stochastic game in remote estimation under dos attacks. *IEEE Control Systems Letters*, 1(1):146–151, 2017.

[37] Yong Ding and Wei Ren. Sampled-data containment control for double-integrator agents with dynamic leaders with nonzero inputs. *Systems & Control Letters*, 139:104673, 2020.

[38] Danny Dolev, Nancy A Lynch, Shlomit S Pinter, Eugene W Stark, and William E Weihl. Reaching approximate agreement in the presence of faults. *Journal of the ACM (JACM)*, 33(3):499–516, 1986.

[39] Xiwang Dong, Yongzhao Hua, Yan Zhou, Zhang Ren, and Yisheng Zhong. Theory and experiment on formation-containment control of multiple multirotor unmanned aerial vehicle systems. *IEEE Transactions on Automation Science and Engineering*, 16(1):229–240, 2018.

[40] GR Duan and RJ Patton. Robust fault detection in linear systems using luenberger observers. In *UKACC International Conference on Control'98 (Conf. Publ. No. 455)*, pages 1468–1473. IET, 1998.

[41] Cynthia Dwork. Differential privacy: A survey of results. In *International conference on theory and applications of models of computation*, pages 1–19. Springer, 2008.

[42] Mohammad Esmalifalak, Ge Shi, Zhu Han, and Lingyang Song. Bad data injection attack and defense in electricity market using game theory study. *IEEE Transactions on Smart Grid*, 4(1):160–169, 2013.

[43] K. Gai, M. Qiu, H. Zhao, and X. Sun. Resource management in sustainable cyber-physical systems using heterogeneous cloud computing. *IEEE Transactions on Sustainable Computing*, PP(99):1–1, 2017.

[44] Robert Gibbons. *A primer in game theory*. Harvester Wheatsheaf, 1992.

[45] Marc Girault, Robert Cohen, and Mireille Campana. A generalized birthday attack. In *Workshop on the Theory and Application of Cryptographic Techniques*, pages 129–156. Springer, 1988.

[46] Zuxing Gu, Hong Song, Yu Jiang, Jeonghone Choi, Hongjiang He, Lui Sha, and Ming Gu. An integrated medical cps for early detection of paroxysmal sympathetic hyperactivity. In *Bioinformatics and Biomedicine (BIBM), 2016 IEEE International Conference In 2016 IEEE International Conference on Bioinformatics and Biomedicine (BIBM)*, pages 818–822. IEEE, 2016.

[47] Hamed Haghshenas, Mohammad Ali Badamchizadeh, and Mahdi Baradarannia. Containment control of heterogeneous linear multi-agent systems. *Automatica*, 54:210–216, 2015.

[48] John Hatcliff, Andrew King, Insup Lee, Alasdair Macdonald, Anura Fernando, Michael Robkin, Eugene Vasserman, Sandy Weininger, and Julian M Goldman. Rationale and architecture principles for medical application platforms. In *Cyber-Physical Systems (ICCPS), 2012 IEEE/ACM Third International Conference In 2012 IEEE/ACM Third International Conference on Cyber-Physical Systems*, pages 3–12. IEEE, 2012.

[49] Jianping He, Lin Cai, Peng Cheng, Jianping Pan, and Ling Shi. Consensus-based data-privacy preserving data aggregation. *IEEE Transactions on Automatic Control*, 64(12):5222–5229, 2019.

[50] Yiguang Hong, Guanrong Chen, and Linda Bushnell. Distributed observers design for leader-following control of multi-agent networks. *Automatica*, 44(3):846–850, 2008.

[51] Wenying Hou, Minyue Fu, Huanshui Zhang, and Zongze Wu. Consensus conditions for general second-order multi-agent systems with communication delay. *Automatica*, 75:293–298, 2017.

[52] He Huang, Yan Lindsay Sun, Qing Yang, Fan Zhang, Xiaorong Zhang, Yuhong Liu, Jin Ren, and Fabian Sierra. Integrating neuromuscular and cyber systems for neural control of artificial legs. In *Proceedings of the 1st ACM/IEEE international conference on cyber-physical systems*, pages 129–138. ACM, 2010.

[53] Ling Huang, Anthony D Joseph, Blaine Nelson, Benjamin IP Rubinstein, and J Doug Tygar. Adversarial machine learning. In *Proceedings of the 4th ACM workshop on Security and artificial intelligence*, pages 43–58, 2011.

[54] Zhenqi Huang, Sayan Mitra, and Geir Dullerud. Differentially private iterative synchronous consensus. In *Proceedings of the 2012 ACM Workshop on Privacy in the Electronic Society*, pages 81–90, 2012.

[55] Zhenqi Huang, Yu Wang, Sayan Mitra, and Geir E Dullerud. On the cost of differential privacy in distributed control systems. In *Proceedings of the 3rd International Conference on High Confidence Networked Systems*, pages 105–114, 2014.

[56] US ICS-CERT. Year in review 2016. 2016.

[57] Marija D Ilic, Le Xie, and Usman A Khan. Modeling future cyber-physical energy systems. In *Power and Energy Society General Meeting-Conversion and Delivery of Electrical Energy in the 21st Century, 2008 IEEE*, pages 1–9. IEEE, 2008.

[58] Meng Ji, Giancarlo Ferrari-Trecate, Magnus Egerstedt, and Annalisa Buffa. Containment control in mobile networks. *IEEE Transactions on Automatic Control*, 53(8):1972–1975, 2008.

[59] Zhihao Jiang, Miroslav Pajic, and Rahul Mangharam. Cyber–physical modeling of implantable cardiac medical devices. *Proceedings of the IEEE*, 100(1):122–137, 2012.

[60] B. Kailkhura, Y. S. Han, S. Brahma, and P. K. Varshney. Distributed bayesian detection in the presence of byzantine data. *IEEE Transactions on Signal Processing*, 63(19):5250–5263, Oct 2015.

[61] Bhavya Kailkhura, Yunghsiang S Han, Swastik Brahma, and Pramod K Varshney. Distributed bayesian detection with byzantine data. *arXiv preprint arXiv:1307.3544*, 2013.

[62] Amir Khazraei, Hamed Kebriaei, and Farzad Rajaei Salmasi. A new watermarking approach for replay attack detection in LQG systems. In *2017 IEEE 56th Annual Conference on Decision and Control (CDC)*, pages 5143–5148. IEEE, 2017.

[63] Amir Khazraei, Hamed Kebriaei, and Farzad Rajaei Salmasi. Replay attack detection in a multi agent system using stability analysis and loss effective watermarking. In *2017 American Control Conference (ACC)*, pages 4778–4783. IEEE, 2017.

[64] Roger M. Kieckhafer and Mohammad H. Azadmanesh. Reaching approximate agreement with mixed-mode faults. *IEEE Transactions on Parallel and Distributed Systems*, 5(1):53–63, 1994.

[65] Jan Kleissl and Yuvraj Agarwal. Cyber-physical energy systems: Focus on smart buildings. In *Proceedings of the 47th Design Automation Conference*, pages 749–754. ACM, 2010.

[66] Chiu-Yuen Koo. Broadcast in radio networks tolerating byzantine adversarial behavior. In *Proceedings of the Twenty-Third annual ACM Symposium on Principles of Distributed Computing*, pages 275–282, 2004.

[67] Prabha Kundur, Neal J Balu, and Mark G Lauby. *Power system stability and control*, volume 7. McGraw-Hill New York, 1994.

[68] David Kushner. The real story of stuxnet. *IEEE Spectrum*, 3(50):48–53, 2013.

[69] Leslie Lamport, Robert Shostak, and Marshall Pease. The byzantine generals problem. *ACM Transactions on Programming Languages and Systems (TOPLAS)*, 4(3):382–401, 1982.

[70] John Kah Soon Lau, Chen-Khong Tham, and Tie Luo. Participatory cyber physical system in public transport application. In *Utility and Cloud Computing (UCC), 2011 Fourth IEEE International Conference In 2011 4th IEEE International Conference on Utility and Cloud Computing*, pages 355–360. IEEE, 2011.

[71] Heath J LeBlanc. *Resilient cooperative control of networked multi-agent systems*. Vanderbilt University, 2012.

[72] Heath J LeBlanc, Haotian Zhang, Xenofon Koutsoukos, and Shreyas Sundaram. Resilient asymptotic consensus in robust networks. *IEEE Journal on Selected Areas in Communications*, 31(4):766–781, 2013.

[73] Heath J LeBlanc, Haotian Zhang, Shreyas Sundaram, and Xenofon Koutsoukos. Consensus of multi-agent networks in the presence of adversaries using only local information. In *Proceedings of the 1st international conference on High Confidence Networked Systems*, pages 1–10, 2012.

[74] Jianzhen Li, Wei Ren, and Shengyuan Xu. Distributed containment control with multiple dynamic leaders for double-integrator dynamics using only position measurements. *IEEE Transactions on Automatic Control*, 57(6):1553–1559, 2011.

[75] Xu Li, Rongxing Lu, Xiaohui Liang, Xuemin Shen, Jiming Chen, and Xiaodong Lin. Smart community: an internet of things application. *IEEE Communications Magazine*, 49(11), 2011.

[76] Yuzhe Li, Ling Shi, Peng Cheng, Jiming Chen, and Daniel E Quevedo. Jamming attacks on remote state estimation in cyber-physical systems: A game-theoretic approach. *IEEE Transactions on Automatic Control*, 60(10):2831–2836, 2015.

[77] Zhongkui Li, Wei Ren, Xiangdong Liu, and Mengyin Fu. Distributed containment control of multi-agent systems with general linear dynamics in the presence of multiple leaders. *International Journal of Robust and Nonlinear Control*, 23(5):534–547, 2013.

[78] Shinyoung Lim, Lawrence Chung, Oakyoung Han, and Jae-Hyun Kim. An interactive cyber-physical system (CPS) for people with disability and frail elderly people. In *Proceedings of the 5th International Conference on Ubiquitous Information Management and Communication*, page 113. ACM, 2011.

[79] Peng Lin and Wei Ren. Constrained consensus in unbalanced networks with communication delays. *IEEE Transactions on Automatic Control*, 59(3):775–781, 2013.

[80] Huiyang Liu, Guangming Xie, and Long Wang. Necessary and sufficient conditions for solving consensus problems of double-integrator dynamics via sampled control. *International Journal of Robust and Nonlinear Control*, 20(15):1706–1722, 2010.

[81] Huiyang Liu, Guangming Xie, and Long Wang. Necessary and sufficient conditions for containment control of networked multi-agent systems. *Automatica*, 48(7):1415–1422, 2012.

[82] Tengfei Liu, Jia Qi, and Zhong-Ping Jiang. Distributed containment control of multi-agent systems with velocity and acceleration saturations. *Automatica*, 117:108992, 2020.

[83] Yao Liu, Peng Ning, and Michael K Reiter. False data injection attacks against state estimation in electric power grids. *ACM Transactions on Information and System Security (TISSEC)*, 14(1):1–33, 2011.

[84] Zhi-Wei Liu, Zhi-Hong Guan, Xuemin Shen, and Gang Feng. Consensus of multi-agent networks with aperiodic sampled communication via impulsive algorithms using position-only measurements. *IEEE Transactions on Automatic Control*, 57(10):2639–2643, 2012.

[85] Sarah M Loos, André Platzer, and Ligia Nistor. Adaptive cruise control: Hybrid, distributed, and now formally verified. In *International Symposium on Formal Methods*, pages 42–56. Springer, 2011.

[86] Youcheng Lou and Yiguang Hong. Target containment control of multi-agent systems with random switching interconnection topologies. *Automatica*, 48(5):879–885, 2012.

[87] An-Yang Lu and Guang-Hong Yang. Secure luenberger-like observers for cyber–physical systems under sparse actuator and sensor attacks. *Automatica*, 98:124–129, 2018.

[88] Yang Lu and Minghui Zhu. A control-theoretic perspective on cyber-physical privacy: Where data privacy meets dynamic systems. *Annual Reviews in Control*, 2019.

[89] Mohammad Hossein Manshaei, Quanyan Zhu, Tansu Alpcan, Tamer Bacşar, and Jean-Pierre Hubaux. Game theory meets network security and privacy. *ACM Computing Surveys (CSUR)*, 45(3):1–39, 2013.

[90] Stefano Marano, Vincenzo Matta, and Lang Tong. Distributed detection in the presence of byzantine attacks. *IEEE Transactions on Signal Processing*, 57(1):16–29, 2008.

[91] Eric Maskin. Nash equilibrium and welfare optimality. *The Review of Economic Studies*, 66(1):23–38, 1999.

[92] Jie Mei, Wei Ren, and Guangfu Ma. Distributed containment control for lagrangian networks with parametric uncertainties under a directed graph. *Automatica*, 48(4):653–659, 2012.

[93] Chiara Mellucci, Prathyush P Menon, Christopher Edwards, and Antonella Ferrara. Second-order sliding mode observers for fault reconstruction in power networks. *IET Control Theory & Applications*, 11(16):2772–2782, 2017.

[94] Hammurabi Mendes and Maurice Herlihy. Multidimensional approximate agreement in byzantine asynchronous systems. In *Proceedings of the Forty-Fifth Annual ACM Symposium on Theory of Computing*, pages 391–400, 2013.

[95] Ziyang Meng, Wei Ren, and Zheng You. Distributed finite-time attitude containment control for multiple rigid bodies. *Automatica*, 46(12):2092–2099, 2010.

[96] Ralph C Merkle. A certified digital signature. In *Conference on the Theory and Application of Cryptology*, pages 218–238. Springer, 1989.

[97] Yilin Mo, João P Hespanha, and Bruno Sinopoli. Resilient detection in the presence of integrity attacks. *IEEE Transactions on Signal Processing*, 62(1):31–43, 2014.

[98] Yilin Mo and Richard M Murray. Privacy preserving average consensus. *IEEE Transactions on Automatic Control*, 62(2):753–765, 2016.

[99] Yilin Mo and Bruno Sinopoli. Secure control against replay attacks. In *2009 47th Annual Allerton Conference on Communication, Control, and Computing (Allerton)*, pages 911–918. IEEE, 2009.

[100] Yilin Mo, Sean Weerakkody, and Bruno Sinopoli. Physical authentication of control systems: Designing watermarked control inputs to detect counterfeit sensor outputs. *IEEE Control Systems*, 35(1):93–109, 2015.

[101] Prashanth Mohan, Venkata N Padmanabhan, and Ramachandran Ramjee. Nericell: rich monitoring of road and traffic conditions using mobile smartphones. In *Proceedings of the 6th ACM Conference on Embedded network Sensor Systems*, pages 323–336. ACM, 2008.

[102] Alcir Monticelli. *State estimation in electric power systems: a generalized approach.* Springer Science & Business Media, 2012.

[103] Angelia Nedic, Asuman Ozdaglar, and Pablo A Parrilo. Constrained consensus and optimization in multi-agent networks. *IEEE Transactions on Automatic Control*, 55(4):922–938, 2010.

[104] Wei Ni and Daizhan Cheng. Leader-following consensus of multi-agent systems under fixed and switching topologies. *Systems & Control Letters*, 59(3-4):209–217, 2010.

[105] Erfan Nozari, Pavankumar Tallapragada, and Jorge Cortés. Differentially private average consensus: Obstructions, trade-offs, and optimal algorithm design. *Automatica*, 81:221–231, 2017.

[106] Reza Olfati-Saber, J Alex Fax, and Richard M Murray. Consensus and cooperation in networked multi-agent systems. *Proceedings of the IEEE*, 95(1):215–233, 2007.

[107] Reza Olfati-Saber and Richard M Murray. Consensus problems in networks of agents with switching topology and time-delays. *IEEE Transactions on Automatic Control*, 49(9):1520–1533, 2004.

[108] Mete Ozay, Inaki Esnaola, Fatos Tunay Yarman Vural, Sanjeev R Kulkarni, and H Vincent Poor. Machine learning methods for attack detection in the smart grid. *IEEE Transactions on Neural Networks and Learning Systems*, 27(8):1773–1786, 2015.

[109] Luca Parolini, Niraj Tolia, Bruno Sinopoli, and Bruce H Krogh. A cyber-physical systems approach to energy management in data centers. In *Proceedings of the 1st ACM/IEEE International Conference on Cyber-Physical Systems*, pages 168–177. ACM, 2010.

[110] Luis A Rademacher. Approximating the centroid is hard. In *Proceedings of the Twenty-Third Annual Symposium on Computational Geometry*, pages 302–305, 2007.

[111] Sandy Rahmé, Yann Labit, and Frédéric Gouaisbaut. An unknown input sliding observer for anomaly detection in TCP/IP networks. In *2009 International Conference on Ultra Modern Telecommunications & Workshops*, pages 1–7. IEEE, 2009.

[112] Gaurav Bhatia Karthik Lakshmanan Ragunathan Raj Rajkumar. An end-to-end integration framework for automotive cyber-physical systems using sysweaver. *AVICPS 2010*, page 23, 2010.

[113] W. Ren. On consensus algorithms for double-integrator dynamics. *IEEE Transactions on Automatic Control*, 53(6):1503–1509, July 2008.

[114] Wei Ren and Randal W Beard. Consensus seeking in multiagent systems under dynamically changing interaction topologies. *IEEE Transactions on Automatic Control*, 50(5):655–661, 2005.

[115] Wei Ren, Randal W Beard, and Ella M Atkins. Information consensus in multivehicle cooperative control. *IEEE Control Systems Magazine*, 27(2):71–82, 2007.

[116] Xiaoqiang Ren and Yilin Mo. Multiple hypothesis testing in adversarial environments: A game-theoretic approach. In *2018 Annual American Control Conference (ACC)*, pages 967–972. IEEE, 2018.

[117] Xiaoqiang Ren, Jiaqi Yan, and Yilin Mo. Binary hypothesis testing with byzantine sensors: Fundamental tradeoff between security and efficiency. *IEEE Transactions on Signal Processing*, 66(6):1454–1468, 2018.

[118] Craig G Rieger, David I Gertman, and Miles A McQueen. Resilient control systems: Next generation design research. In *2009 2nd Conference on Human System Interactions*, pages 632–636. IEEE, 2009.

[119] Helena Rifa-Pous and Jordi Herrera-Joancomartí. Computational and energy costs of cryptographic algorithms on handheld devices. *Future Internet*, 3(1):31–48, 2011.

[120] Cynthia Rudin, David Waltz, Roger N Anderson, Albert Boulanger, Ansaf Salleb-Aouissi, Maggie Chow, Haimonti Dutta, Philip N Gross, Bert Huang, Steve Ierome, et al. Machine learning for the New York city power grid. *IEEE Transactions on Pattern Analysis and Machine Intelligence*, 34(2):328–345, 2011.

[121] Alessandra Sala, Xiaohan Zhao, Christo Wilson, Haitao Zheng, and Ben Y Zhao. Sharing graphs using differentially private graph models. In *Proceedings of the 2011 ACM SIGCOMM Conference on Internet Measurement Conference*, pages 81–98, 2011.

[122] Henrik Sandberg, Saurabh Amin, and Karl Henrik Johansson. Cyber-physical security in networked control systems: An introduction to the issue. *IEEE Control Systems Magazine*, 35(1):20–23, 2015.

[123] Henrik Sandberg, György Dán, and Ragnar Thobaben. Differentially private state estimation in distribution networks with smart meters. In *2015 54th IEEE Conference on Decision and Control (CDC)*, pages 4492–4498. IEEE, 2015.

[124] UWE SCHMOCK. Large deviations techniques and applications. *Journal of the American Statistical Association*, 95(452):1380–1380, 2000.

[125] Eugene Seneta. Coefficients of ergodicity: Structure and applications. *Advances in Applied Probability*, 11(3):576–590, 1979.

[126] Claude E Shannon. Communication theory of secrecy systems. *Bell Labs Technical Journal*, 28(4):656–715, 1949.

[127] Stephen Smaldone, Chetan Tonde, Vancheswaran K Ananthanarayanan, Ahmed Elgammal, and Liviu Iftode. The cyber-physical bike: A step towards safer green transportation. In *Proceedings of the 12th Workshop on Mobile Computing Systems and Applications*, pages 56–61. ACM, 2011.

[128] E. Soltanmohammadi, M. Orooji, and M. Naraghi-Pour. Decentralized hypothesis testing in wireless sensor networks in the presence of misbehaving nodes. *IEEE Transactions on Information Forensics and Security*, 8(1):205–215, Jan 2013.

[129] Lili Su and Nitin H Vaidya. Fault-tolerant distributed optimization (part iv): Constrained optimization with arbitrary directed networks. *arXiv preprint arXiv:1511.01821*, 2015.

[130] Shreyas Sundaram and Bahman Gharesifard. Distributed optimization under adversarial nodes. *IEEE Transactions on Automatic Control*, 2018.

[131] Bilal Syed, Arpan Pal, Krishnan Srinivasarengan, and P Balamuralidhar. A smart transport application of cyber-physical systems: Road surface monitoring with mobile devices. In *Sensing Technology (ICST), 2012 Sixth International Conference In 2012 6th International Conference on Sensing Technology (ICST)*, pages 8–12. IEEE, 2012.

[132] Anastasios Tsiamis, Konstantinos Gatsis, and George J Pappas. State estimation with secrecy against eavesdroppers. *IFAC-PapersOnLine*, 50(1):8385–8392, 2017.

[133] J. Usevitch and D. Panagou. Resilient leader-follower consensus to arbitrary reference values in time-varying graphs. *IEEE Transactions on Automatic Control*, 65(4), pp.1755–1762. 2019.

[134] James Usevitch and Dimitra Panagou. Resilient leader-follower consensus to arbitrary reference values in time-varying graphs. *IEEE Transactions on Automatic Control*, 2019.

[135] Nitin H Vaidya. Iterative byzantine vector consensus in incomplete graphs. In *International Conference on Distributed Computing and Networking*, pages 14–28. Springer, 2014.

[136] Nitin H Vaidya and Vijay K Garg. Byzantine vector consensus in complete graphs. In *Proceedings of the 2013 ACM Symposium on Principles of Distributed Computing*, pages 65–73, 2013.

[137] Nitin H Vaidya, Lewis Tseng, and Guanfeng Liang. Iterative approximate byzantine consensus in arbitrary directed graphs. In *Proceedings of the 2012 ACM Symposium on Principles of Distributed Computing*, pages 365–374. ACM, 2012.

[138] Kyriakos G Vamvoudakis, Joao P Hespanha, Bruno Sinopoli, and Yilin Mo. Detection in adversarial environments. *IEEE Transactions on Automatic Control*, 59(12):3209–3223, 2014.

[139] Xin Wang, Hideaki Ishii, Linkang Du, Peng Cheng, and Jiming Chen. Privacy-preserving distributed machine learning via local randomization and admm perturbation. *arXiv preprint arXiv:1908.01059*, 2019.

[140] Xuan Wang, Shaoshuai Mou, and Shreyas Sundaram. A resilient convex combination for consensus-based distributed algorithms, 2018.

[141] Yu Wang, Zhenqi Huang, Sayan Mitra, and Geir E Dullerud. Differential privacy in linear distributed control systems: Entropy minimizing mechanisms and performance tradeoffs. *IEEE Transactions on Control of Network Systems*, 4(1):118–130, 2017.

[142] Ermin Wei and Asuman Ozdaglar. Distributed alternating direction method of multipliers. In *Decision and Control (CDC), 2012 IEEE 51st Annual Conference on*, pages 5445–5450. IEEE, 2012.

[143] Wei Wu, Muhammad Khalid Aziz, Hantao Huang, Hao Yu, and Hoay Beng Gooi. A real-time cyber-physical energy management system for smart houses. In *Innovative Smart Grid Technologies Asia (ISGT), 2011 IEEE PES*, pages 1–8. IEEE, 2011.

[144] Feng Xia and Jianhua Ma. Building smart communities with cyber-physical systems. In *Proceedings of 1st international symposium on From digital footprints to social and community intelligence*, pages 1–6. ACM, 2011.

[145] W. Xia, J. Liu, M. Cao, K. H. Johansson, and T. Başar. Generalized sarymsakov matrices. *IEEE Transactions on Automatic Control*, 64(8):3085–3100, Aug 2019.

[146] Lin Xiao, Stephen Boyd, and Seung-Jean Kim. Distributed average consensus with least-mean-square deviation. *Journal of Parallel and Distributed Computing*, 67(1):33–46, 2007.

[147] Quan Xiong, Peng Lin, Zhiyong Chen, Wei Ren, and Weihua Gui. Distributed containment control for first-order and second-order multiagent systems with arbitrarily bounded delays. *International Journal of Robust and Nonlinear Control*, 29(4):1122–1131, 2019.

[148] Jiaqi Yan, Fanghong Guo, and Changyun Wen. Attack detection and isolation for distributed load shedding algorithm in microgrid systems. *IEEE Journal of Emerging and Selected Topics in Industrial Electronics*, 1(1):102–110, 2020.

[149] Jiaqi Yan, Fanghong Guo, and Changyun Wen. False data injection against state estimation in power systems with multiple cooperative attackers. *ISA transactions*, 101:225–233, 2020.

[150] Jiaqi Yan, Xiuxian Li, Yilin Mo, and Changyun Wen. Resilient multi-dimensional consensus and optimization in adversarial environment. *arXiv preprint arXiv:2001.00937*, 2020.

[151] Jiaqi Yan, Yilin Mo, Xiuxian Li, and Changyun Wen. A "safe kernel" approach for resilient multi-dimensional consensus. *arXiv preprint arXiv:1911.10836*, 2019.

[152] Jiaqi Yan, Xiaoqiang Ren, and Yilin Mo. Sequential detection in adversarial environment. In *Decision and Control (CDC), 2017 IEEE 56th Annual Conference In 2017 IEEE 56th Annual Conference on Decision and Control (CDC)*, pages 170–175. IEEE, 2017.

[153] Jiaqi Yan and Changyun Wen. Resilient containment control in adversarial environment. *IEEE Transactions on Control of Network Systems*, 7(4):1951–1959, 2020.

[154] Jiaqi Yan, Changyun Wen, Xiao-Kang Liu, and Lantao Xing. Resilient impulsive control for second-order consensus under malicious nodes. *IEEE Transactions on Circuits and Systems II: Express Briefs*, 68(6):1962–1966, 2020.

[155] Qingyu Yang, Dou An, Rui Min, Wei Yu, Xinyu Yang, and Wei Zhao. On optimal PMU placement-based defense against data integrity attacks in smart grid. *IEEE Transactions on Information Forensics and Security*, 12(7):1735–1750, 2017.

[156] Maojiao Ye, Guoqiang Hu, and Shengyuan Xu. An extremum seeking-based approach for Nash equilibrium seeking in N-cluster noncooperative games. *Automatica*, 114:108815, 2020.

[157] Wenwu Yu, Guanrong Chen, and Ming Cao. Some necessary and sufficient conditions for second-order consensus in multi-agent dynamical systems. *Automatica*, 46(6):1089–1095, 2010.

[158] Wenwu Yu, Wei Ren, Wei Xing Zheng, Guanrong Chen, and Jinhu Lü. Distributed control gains design for consensus in multi-agent systems with second-order nonlinear dynamics. *Automatica*, 49(7):2107–2115, 2013.

[159] Fumin Zhang and Zhenwu Shi. Optimal and adaptive battery discharge strategies for cyber-physical systems. In *Decision and Control, 2009 Held Jointly with the 2009 28th Chinese Control Conference. CDC/CCC 2009. Proceedings of the 48th IEEE Conference In Proceedings of the 48th IEEE Conference on Decision and Control (CDC) held jointly with 2009 28th Chinese Control Conference*, pages 6232–6237. IEEE, 2009.

[160] Fumin Zhang, Zhenwu Shi, and Wayne Wolf. A dynamic battery model for co-design in cyber-physical systems. In *Distributed Computing Systems Workshops, 2009. ICDCS Workshops' 09. 29th IEEE International Conference In 2009 29th IEEE International Conference on Distributed Computing Systems Workshops*, pages 51–56. IEEE, 2009.

[161] Chengcheng Zhao, Jianping He, and Qing-Guo Wang. Resilient distributed optimization algorithm against adversarial attacks. *IEEE Transactions on Automatic Control*, 2019.

[162] Wei Zhu and Daizhan Cheng. Leader-following consensus of second-order agents with multiple time-varying delays. *Automatica*, 46(12):1994–1999, 2010.

For Product Safety Concerns and Information please contact our EU representative GPSR@taylorandfrancis.com Taylor & Francis Verlag GmbH, Kaufingerstraße 24, 80331 München, Germany